山东省优势特色学科（建筑学）
山东省研究生教育优质课程"城市设计理论"资助出版

城乡规划研究生专业英语

——可持续城市空间与设计

An English Coursebook for Urban and
Rural Planning Postgraduates
—Sustainable City Space and Design

杨慧 赵亮 侯慧心 ◎ 编著

天津大学 出版社
TIANJIN UNIVERSITY PRESS

图书在版编目（CIP）数据

城乡规划研究生专业英语：可持续城市空间与设计
山东省优势特色学科（建筑学）、山东省研究生教育优
质课程"城市设计理论" / 杨慧, 赵亮, 侯慧心编著
. -- 天津：天津大学出版社, 2022.7
ISBN 978-7-5618-7254-3

Ⅰ. ①城⋯ Ⅱ. ①杨⋯ ②赵⋯ ③侯⋯ Ⅲ. ①城乡规
划—英语—研究生—教材 Ⅳ. ① TU98

中国版本图书馆 CIP 数据核字（2022）第 133388 号

出版发行	天津大学出版社	
地　　址	天津市卫津路 92 号天津大学内（邮编：300072）	
电　　话	发行部：022-27403647	
网　　址	publish.tju.edu.cn	
印　　刷	北京盛通商印快线网络科技有限公司	
经　　销	全国各地新华书店	
开　　本	787 毫米 × 1092 毫米　1/16	
印　　张	14.25	
字　　数	405 千	
版　　次	2022 年 11 月第 1 版	
印　　次	2022 年 11 月第 1 次	
定　　价	59.00 元	

前言 | Preface

本书是国内第一本专门应用于城乡规划专业研究生培养的专业英语教材。英语理解与翻译已是研究生的必需技能，"研究生专业英语"课程是学位课、必修课，相关教材不仅应在本课程中使用，还应成为服务研究生开展学习与科研工作的参考书、案例源。本书的编写目的即是适用多课程、面向多学科、覆盖更高层次的研究生学习、贯通整个培养体系。

本书共设置 6 个单元，包含"导读—预备知识—课文讲解—词汇—练习与思考—课后延伸" 6 个教学任务，便于灵活对应 24 课时的教学内容。各单元以研究专题的形式呈现，充分结合城市规划学科的研究方向，可与"城市与区域规划研究""现代城市规划理论""城市设计理论""城市住房政策"等课程共享知识模块，不仅为学生提供专业英语优质教材，也为其他课程提供城乡规划案例资源与研究素材，实现课程之间的教学互动，以及"课程教学 + 研究方向"的双向协同。

从选题来看，本书以"可持续城市空间与设计"为主题，涉及"城市发展模式""住房与社区""城市韧性""气候适应性设计""碳达峰碳中和""健康城市""全民友好"等热点议题。特别说明一点，本书第 1 课以"中国营城智慧"开篇，回溯中国传统城市建设中的匠心巧思，希望读者在语言文字中感受国家的发展与文化的自信。

从选材来看，本书坚持选择典型案例、面向地域空间。如"城市与区域发展"专题中选择伦敦为典型案例，英国是现代城市规划的起源，也是城乡规划学科的起点。又如在"可持续发展"专题中选择的是亚洲国家——日本与中国——进行讨论，内容包括对 2035 年碳达峰碳中和目标、可持续发展指数等关键指标的分析，这些都是面向地域可持续发展的研究内容。

同时，本书重视选材质量与时效。伦敦案例来自 2021 年 3 月公布的《伦敦规划 2021》（*London Plan 2021*），是伦敦的总体空间发展战略，是未来 20 年伦敦城市发展的综合框架，也是遭受疫情肆虐后的城市思考。日本案例来源于 2021 年 7 月麦肯锡咨询公司（Mckinsey & Compony）公开发布的研究报告 *How Japan Could Reach Carbon Neutrality by 2050*。

从结构来看，本书采用"左英右中"对页对照的形式，更加直观，便于读者对照学习。但这种对照方式对译文质量要求更高，也极大增加了排版难度，在此特别感谢天津大学出版社刘大馨、杨晔等编辑的辛苦付出。

另外还要特别感谢世界资源研究所中国可持续城市部门主任刘岱宗先生及其创办的"一览众山小——可持续城市与交通"微信公众平台的所有志愿者，特别是陈航、杨潇晗、黎赟、邱馨仪、张鹤鸣、王琛芳、张浩然、赵柄智、冷怡霖、张佳玥、陈俊豪、杨蕊源、林煜、余铭航、吴梦洋、陆建、林若然，他们对本书的选题与组稿提供了帮助和支持。

感谢山东建筑大学建筑城规学院2019级范亭君同学，她在本书编写过程中做了大量工作。

感谢山东省优势特色学科（建筑学）、山东省研究生教育优质课程"城市设计理论"资助出版。

本书结构

课次	专题	知识点	案例	文献来源及性质
1	中国营城智慧 Chinese Wisdom of Urban Construction	庙宇般的城市 The City a Temple	北京 Beijing	[图书] 《城镇与建筑》 *Towns and Buildings*
		古今北京 The Traditional and Modern Beijing	北京 Beijing	[图书] 《建筑原理：形式材料的基本原则》 *Architecture Principles:Architecture Principles Form*
2	城市与区域发展 Urban and Regional Development	美好增长 Good Growth	伦敦 London	[城市总体规划] 《伦敦规划 2021》 *London Plan 2021*
		区域协同 Regional Collaboration	伦敦 London	[规划实践] 《区域协同·发展潜力地区》 *Regional Collaboration · Development of Potential Areas*
3	可持续发展 Sustainable Development	碳达峰碳中和目标 The Carbon Peak and Carbon Neutrality Target	日本、中国 Japan, China	[研究报告] 《日本 2050 年碳中和之路》 *How Japan Could Reach Carbon Neutrality by 2050*
		碳成本临界点 The Carbon Tipping Point of Cost	亚洲国家 Asian Countries	[研究报告] 《基础设施的碳成本——气候》 危机的关键 *Carbon Cost in Infrastructure—the Key to the Climate Crisis*
4	城市交通 Urban Transport	城市移动性 Urban Mobility	纽约、米兰、莫斯科、首尔、伦敦、香港 New York, Milan, Moscow, Seoul, London, Hong Kong	[研究报告] 《成功的要素——2021 全球 25 城交通报告》 *Elements of Success—Urban Transportation Systems of 25 Global Cities 2021*
		交通政策 Transport Policy	北卡罗来纳 North Carolina	[评论文章] 《"慢街"政策与城市规划》 *Slow Street Policy and Urban Planning*

续表

课次	专题	知识点	案例	文献来源及性质
5	城市韧性 City Resilience	城市防灾 Urban Disaster Prevention	瑞士 Swiss	[规划实践] 《城市气候循环行动手册》 *Urban Action-Going Circular Booklet*
		气候适应性设计 Climate Adaptive Design	瑞士 Swiss	[规划实践] 《建立韧性》 *Building Resilience*
6	城市社区与公共空间 Urban Community and Public Space	城市健康与人民福祉 Urban Health and People's Wellbeing	加拿大 Canada	[评论文章] 《应对压力、保持活力：以城市公共空间的方式》 *How to Make the Most of Covid Winter*
		住房与社区 Housing and Community	伦敦 London	[规划实践] 《建设强大和包容的社区》 *Building Strong and Inclusive Communities*
附录 1	城市规划术语 Technical Terms of Urban Planning			
附录 2	可持续城市空间与设计 Sustainable Urban Space and Design			

如何使用本书

本教材在各专题中采用"导读—预备知识—课文讲解—课后延伸"的形式展开，中英对照，由浅入深，教师可以依据学生能力选择进程，避免"吃不饱"与"咽不下"的学习差距。

本教材已将每个研究专题的内容进行了模块化预拆解，如"导读"和"预备知识"，为实施"翻转课堂"等互动式教学方法提供了极大便利，有助于使用本教材的高校进行相关教学方法的尝试，培养学生的主动学习习惯与专业英语应用能力。

目录 | Contents

第 1 课　中国营城智慧

Lesson 1　Chinese Wisdom of Urban Construction

导读 ｜ Introduction

别样的北京城 ｜ A Different City of Beijing

　　世界著名建筑师、城市规划师斯坦·埃勒·拉斯穆森（Steen Eiler Rasmussen）在其著作《城镇与建筑》（*Towns and Buildings*）中提到，"基本上德国和日本都出版了很好的导游书籍，介绍了北京每一座宫殿和寺庙的详细信息，但是对北京城本身却没有特别介绍。其实北京是世界上的城市建设奇迹之一，该城对称布局，是一座独特的、不朽的都城，显示了当时极高的文明——这是要我们亲身体会才能知道的"。

　　按照城市建设主题和特征，拉斯穆森在全球选择了 16 个城市进行分析和说明，他将北京城放在著作的首篇，赋予其"庙宇般的城市"（The City a Temple）的主题。将北京放在名家名作的开篇，说明西方学者认可北京的城市规划与建设成果，也证明了中国古代城市规划在世界城市规划历史中的重要地位，亦代表了西方科学家拥有客观公正的研究态度，更增强了我们的文化自信。

　　世界上没有两座城市的设计是完全一样的，城市表达了一定的理想，也蕴藏着一段历史。

预备知识 | Preliminary Knowledge

中国城市建设史及文化遗产术语 | Technical Terms for History of Chinese Urban Construction and Cultural Heritage

1. 聚（ju），settlement。中国古代自发形成的聚居地。其是否采用规整形态作为居住组织方式以及规模是否较小仍不明确，但与后世的村庄形态有较大差异。

2. 邑（yi），settlement。中国古代规制聚落的统称。本义为封地，最初"国""都"均称邑，后特指一般规制型聚落。通常采用规整的里作为居住组织方式。

3. 都（du），capital。本义为有已故君主宗庙的邑，后引申为具有重要政治地位的邑。

4. 国（guo），capital。天子或诸侯的都城。后转指封疆，"国"的本义已被"都"取代。

5. 鄙（bi），frontier settlement。地处边远地区的邑。

6. 城（cheng），city wall，city。（1）都邑四周的墙垣；（2）聚落的通称，即城市。

7. 市（shi），market。中国古代对市场、集市的简称。

8. 郭（guo），outer city wall。又称"廓"。外城，古代在城的外围加筑的一道城墙。

9. 城邑（chengyi），walled city。有城垣的邑。

10. 城池（chengchi），city fortress。（1）由城郭与池隍构成的中国古代城居聚落防御工程；（2）代指城市。

11. 城厢（chengxiang），old town area。俗称"老城厢"。城墙以内建成区（城）和城墙以外建成区（厢）的合称，古代泛指城市整体，现代特指城墙内外的旧城区。

12. 水关（shuiguan），water gate。城市水门附近设置的关卡，通常为关厢商埠繁盛之处。

13. 宫城（gongcheng），palace city，forbidden city。都城内由城垣围绕供帝王居住的宫殿区。

14. 皇城（huangcheng），imperial city。都城内由城垣围绕的以宗庙、官署、苑囿为主的地域。

15. 王城（wangcheng），royal city。天子之都城，特指周王所住的洛邑。

16. 苑囿（yuanyou），royal woods。古代帝王豢养禽兽以供游猎的封闭式园林。

17. 离宫（ligong），alternative royal residence。位于都城之外的永久性宫殿，如秦阿房宫、汉建章宫、唐华清宫。

18. 行宫（xinggong），palace for short stays。古代帝王出巡所住的临时居所。

19. 京畿（jingji），capital and environs。（1）国都及其周边地区；（2）引申为国都所在的一级行政区的泛称，如秦之"内史"，汉之"京兆""司隶"，唐之"京畿道"，宋之"京畿路"，明清之"直隶"。

20. 王畿（wangji），royal city area。天子直接管辖统治的疆域，区别于诸侯封国。

21. 满城（mancheng），manchou quarter。清代城市中专供旗人生活居住的城中之城。

22. 埠（bu），quayside，又称"埗"。（1）原指水边、码头；（2）后指古代城市中水边商船集结交易的商贸区。

23. 商埠（shangbu），commercial area。（1）埠原作"埗"，原指水边、码头，因常有商船集结交易形成商贸区，故称"商埠"；（2）近代转指对外开放城市的租界区。

24. 市镇（shizhen），town。宋代以后形成的非农从业人口集中的聚落，为农村的商业贸易中心或手工业集中地。

25. 市井（shijing），market and living area。（1）古代城市中集中交易的场所；（2）现多指市民社会。

26. 草市（caoshi），rural market。中国古代在城市以外地区自发形成的定期集市，是市镇的早期形态。

27. 御街（yujie），royal street。古代国都城中帝王出行的大街，常为城市布局中轴。

28. 阁道（gedao），aisle。古代城市中连接宫殿楼阁的通道。因两侧有廊，廊上开窗如阁，故名"阁道"。

29. 驰道（chidao），rapid road。古代都城和主要城市之间供车辆通行的道路。

30. 里坊（lifang），walled neighborhood and block，又称"坊里""闾里"。城市中有垣墙围绕的居民区。

31. 番坊（fanfang），foreigner block。城市中供外国商人居住的地区。

32. 营国制度（yingguo zhidu），yingguo system。周代关于王城的营建制度，奠定了方形城制、宫城居中、对称布局、礼制等级等中国城市建设传统。

33. 坊巷制（fangxiangzhi），fangxiang sytem，system of blocks and lanes。宋代以后的城市中，将坊里施加于街巷制布局的管理组织方式。

34. 多京制（duojingzhi），multi-capital system。中国古代两个或以上城市具有都城或准都城地位的制度。

35. 国野制（guoyezhi），guo-ye system。又称"乡遂制"。西周时期对直辖区域和属地进行差异化组织管理的制度。

36. 井田制（jingtianzhi），jingtian system，ancient system of farming land subdivision。先秦时期将土地划分为井形方田并分配给农户耕作的土地分配制度。方里之地以井字形划分为九块，外围八块为私田，中间一块为公田。

37. 择中论（zezhonglun），zezhong theory，central location theory。中国古代在城址选择和城中宫室布置等方面所采取的将显要功能置于中心区位的规划布局原则。

38. 因势论（yinshilun），yinshi theory，situation adaptive theory。中国古代城市中因应地形、地势条件规划布局的原则。

39. 社稷（sheji），regime。（1）土地神（社）和谷神（稷）的合称，中国古代城市国家祭祀活动的建筑；（2）后世指代国家政权。

40. 形胜（xingsheng），location advantage。（1）古代指地理区位与自然环境优势；（2）尤指地势险要或有重要战略地位。

41. 风水（fengshui），geometric omen，又称"堪舆""相土"。中国古代占相阳宅（城邑、住宅）、阴宅（墓葬）形胜以卜吉凶的方术。

42. 平江图（Pingjiangtu），map of Pingjiang City。南宋绍定二年（1229年）碑刻平江府城平面图，反映了坊市重整后平江府城的格局，重点描绘了城池、河道、街巷、桥梁、衙署、寺观等，是研究宋代坊里制向坊巷制转变的重要史料。

43. 陵邑（lingyi），mausoleum village。中国古代设置的守卫帝王陵园的城邑。

44. 卫所（weisuo），fortress。中国古代军民分籍制下安置军户驻屯形成的军事组织及其城市，包括卫与所两种。

45. 村寨（cunzhai），village。（1）简称"寨"，择险而守的防卫型村落，通常具有木栅、营垒等防守设施；（2）转指边缘山区少数民族村落，《正韵》"藩落也"，如苗寨。

46. 礼制建筑（lizhi jianzhu），lizhi building。中国古代社会具有宗法礼教功能的建筑，包括宗庙、社稷、明堂、辟雍等。

47. 合院建筑（heyuan jianzhu）, courtyard house。历史上形成的围绕院落空间进行布置的建筑组群，主要有四合院、三合院等不同院落形式。

48. 民居（minju）, vernacular dwelling。普通百姓的居住房屋。由于地理气候条件和生活方式不相同，在不同地域形成形式多样的居住房屋样式和风格。

49. 胡同（hutong）, lane。北京及北方城市的传统风貌街巷，多为四合院等院落住宅构成的街道和巷道。

50. 里弄（lilong）, a kind of residential form with hierarchical streets。俗称"石库门"。近代开埠后出现在上海、天津、汉口等城市的聚居点基本单元，里弄布局紧凑，有主弄、次弄、支弄之分。

51. 租界（zujie）, concession。一国通过条约规定在别国领土上设置的具有行政自治权和治外法权的居留地。

52. 文化遗产（cultural heritage）。人类社会发展历程中留存下来的具有历史、科学和艺术价值的文物和传统文化，也是市民集体记忆的表现和不可再生的文化资源，可分为物质文化遗产和非物质文化遗产。

53.《保护世界文化和自然遗产公约》（*Convention Concerning the Protection of the World Cultural and Natural Heritage*）。简称《世界遗产公约》，是 1972 年 11 月 16 日在巴黎举行的第十七届联合国教科文组织大会中通过的保护对全人类具有突出的普遍价值的文化和自然遗产的公约。

54. 建筑遗产（architectural heritage）。纪念性建筑和位于历史城市村镇中的次要建筑群及其自然环境和人工环境。建筑遗产在提供生活环境品质、维持社会和谐平衡以及文化教育领域方面发挥着重要作用。

55. 文化景观（cultural landscape）。人和自然多样互动形成的景观。根据特征分为三类。①人类主动设计的景观，包括庭院和公园等，美学和实用往往是其重要的建造原因，这些景观有时会和宗教或其他古迹关联。②有机进化的景观，是人类社会、经济、管理、宗教作用形成的结果，是对其所在自然环境顺应和适应的结果。③关联和联想的文化景观，其重点在于自然元素在宗教、艺术和文化上的强烈联系，而文化上的物质实证退居到次要地位。

56. 文化线路（cultural route）。遗产的形态特征定型和形成基于其自身的动态发展和功能演变，展示人类迁徙和交流的特殊的文化现象，代表了一定时间内国家和地区内部或国家和地区之间人的交往和文化传播的陆上道路、水路或者混合类型的通道。

57. 真实性（authenticity），又称"原真性"。即遗产价值的信息来源真实可靠。对与文化遗产的最初与后续特征有关的信息来源及其意义的认识与了解是全面评估真实性的必备基础。这些来源可包括形式与设计、材料与物质、用途与功能、传统与技术、地点与背景、精神与感情以及其他内在或外在因素。

58. 完整性（integrity）。完整性用来衡量自然和/或文化遗产及其特征的整体性和无缺憾性。需要评估遗产符合以下特征的程度：①包括所有表现其突出普遍价值的必要因素；②面积足够大，确保能完整地体现遗产价值的特色和过程；③受到发展的负面影响和/或缺乏维护。

59. 突出的普遍价值（outstanding universal value, OUV）。世界遗产的文化和/或自然价值对全人类的现在和未来所具有的普遍的重大意义。

60. 评估（assessment）。依据对文物古迹及相关历史、文化的调查研究，对文物古迹的价值、保存状况和管理条件作出的科学评价。

61. 价值评估（value assessment）。根据对遗产及相关历史、文化的调查、研究，对遗产的历史价值、艺术价值、科学价值以及社会价值和文化价值、保存状况和管理条件作出的评价。

62. 价值阐释（interpretation of values）。为了提高公众的文化遗产意识和增强文化遗产理解的所有活动，包括出版、讲座、装置、教育项目、社区活动等以及对阐释过程本身的研究、培训和评估。

63. 修复（restoration）。为精准地展现文物古迹或历史建筑某一特定历史时期的形式、外观及特征，通过去除历史上其他时期添加的部件，恢复该时代的缺失部分的行为或做法。

64. 保护（conservation）。对保护项目及其环境进行的科学的调查、勘测、评估、登录、修缮、维修、改善、利用的过程。

65. 就地保护（in situ protection），在文物古迹和历史建筑的原址和原环境中开展保护和改善的活动。

66. 易地保护（out-of-spot protection），当文物古迹和历史建筑面临不可抗拒的因素影响，如自然灾害、水库建设等，通过工程措施，将其空间位移按原样重建安置的措施。

67. 分类保护（classified protection），针对文化遗产不同类别的特点，采取有针对性的保护措施的基本原则，特别强调要关注文物古迹、历史建筑、历史文化街区、历史文化名城保护方法的差异性。

68. 最小干预（minimal intervention），又称"最低限度干预"。以保证文物古迹安全为前提，避免过度干预影响文物古迹价值及历史文化信息的保护方式。

69. 可逆性（reversibility）。对文物古迹所采取的必要干预和改变，由于历史信息和技术条件的不确定性或局限性，宜采取在一定阶段后可以恢复到干预前状态的措施处理。

70. 城市保护（urban conservation）。保护城市中的文物古迹、历史街区，保护和延续古城的传统格局和风貌特色，继承和发扬优秀历史文化传统，保护非物质文化遗产，如民间艺术、传统工艺、传统戏曲、民俗精华、节日活动等。

71. 历史性城市（镇）景观（historic urban landscape, HUL）。文化和自然价值及属性在历史上层层积淀而产生的城市（镇）区域。其超越了"历史中心"或"整体"的概念，包括更广泛的城市（镇）背景及其地理环境。

72. 历史城区（historic urban area）。（1）城镇中能体现其历史发展过程或某一发展时期风貌的地区，涵盖一般通称的古城区和老城区；（2）特指历史文化名城中历史范围清楚、格局和风貌保存较为完整、需要整体保护的地区。

73. 历史文化街区（historic conservation area）。经省、自治区、直辖市人民政府核定公布的保存文物特别丰富、历史建筑集中成片、能够较完整和真实地体现传统格局和历史风貌，并具有一定规模的历史地段。

74. 历史地段（historic area）。能够真实地反映一定历史时期传统风貌和民族、地方特色的地区。

75. 历史风貌（historic landscape）。反映历史文化特征的城镇、乡村景观和自然、人文环境的整体面貌。

（文献来源：城乡规划学名词审定委员会. 城乡规划学名词. 北京：科学出版社，2021）

课文讲解 | Text Explanation

The City a Temple

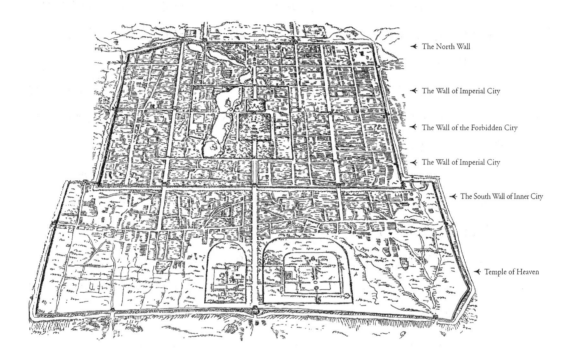

The North Wall

The Wall of Imperial City

The Wall of the Forbidden City

The Wall of Imperial City

The South Wall of Inner City

Temple of Heaven

1 Peking, the capital of China! Has there ever been a more majestic and **illuminative**[1] example of sustained town-planning?

2 It was a city of a million inhabitants but quite different from our idea of a **metropolis**[2]. For miles on end the living-quarters consisted of grey, **one-storeyed**[3] houses lying along narrow, dusty-grey roads behind walls over which rose the tops of green trees. It was like a village, but a village out of all proportion—three miles in one direction and five miles in the other. Yet **coexistent**[4] with the village-like aspect of the **residential**[5] sections there was a **grandeur**[6] in the lay-out of the entire city not to be found in any European capital. Following a clear principle, straight highways, broader than the Paris boulevards, run through the whole town.

庙宇般的城市

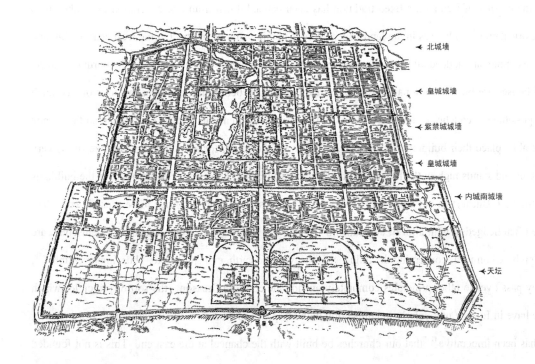

← 北城墙

← 皇城城墙

← 紫禁城城墙

← 皇城城墙

← 内城南城墙

← 天坛

北京，这座中国古都！人类有史以来从未有过如此壮丽辉煌的城市规划先例。

这是一座拥有百万居民的都市，但与我们概念中的大都市却完全不同。住宅区里散布着绵延数英里（注：1 英里 ≈ 1.6 千米）的狭窄胡同，小巷是土灰色的，两旁全是灰色的单层住屋，庭园里遍植花木，绿色的树顶探出墙外，颇有乡村风味，但比起乡村来却又大得太多了。城的一侧有 3 英里长，另一侧有 5 英里长，与乡村风貌住宅区同时共存的是全城雍容堂皇的另一面。比巴黎的林荫道还要宽阔的笔直道路贯穿全城，令欧洲各国的首都望而却步。

3 Peking is built up according to a system of rules which, to a European, seems half mysticism, half common-sense. But neither of these terms is really adequate; the ideas and convictions [7] of a culture like the Chinese can never be completely explained by words derived from a culture like ours.

4 In many parts of Denmark a fixed tradition has been handed down from one generation to another as to the orientation [8] of houses in the landscape. There is nothing strange in this. On the west coast of Jutland, for example, and indeed all the way down along the coast of the North Sea right into the north of France, all houses are built in long parallel wings, squatting down [9] behind the dune [10] rows to avoid as much as possible the clutches of the fierce [11] west wind. It is just as sensible that the Chinese, in so far as they are able, place their buildings so that they open toward the south; for in a climate where the sun is very potent and stands high in the heavens a southern exposure with large, projecting roofs over the buildings is the best. But just as old home remedies—which on closer examination prove to be sensible enough— are often hedged in with a lot of abracadabra [12], Chinese building principles, sensible in themselves, are often based on ideas concerning the influence of heaven and earth, evil spirits, etc. Thus, imperceptibly [13], they pass over to other principles completely outside the control of reason—the sort of principles which we have in Europe particularly in connection with temples and churches. Since early times, for instance, it has been imperative [14] that our churches be built with the chancel at the east end. This is not founded on commonsense. It is ritual [15]. For the Chinese all building practice was, really, ritual, and if there were no established rules for a given case, the priests [16], who were deemed to be in intimate contact with the powers that govern nature, would have to be consulted. And when every house and every temple had to be built according to ritual, how much more important this must have been for the laying out of the main city of the entire empire! For Peking was much more than a capital. It was the residence of the emperor and the emperor was more than a regent [17] or a sovereign [18]; he was a demigod, called "the Son of Heaven", and had the functions of supreme pontiff [19]. Every year at midwinter it was his prerogative [20] to make the great sacrifices [21] to heaven which established the pact between man and the Omnipotent [22] and insured a good year. He succeeded in making himself the spiritual head of his people. His throne [23] was sanctuary, the throne hall a temple facing due south, the entire city a temple ground.

北京是按照一套系统的设计规则逐步兴建而成的，照欧洲人的眼光来看，其中既有神秘主义的内涵，又由常识惯例所成就，但这些解释语汇均不恰当。如用欧洲文化来衡量中国的思想观念，肯定无法求得适切完整的解答。

在丹麦的很多地区，有一种世代相传的惯例，就是在建屋时，先在空地上确定房屋的方位。在这里，这一景象司空见惯，例如从日德兰（Jutland）的西海岸沿北海海岸线，向右进入法国北部，那一带的房屋都筑有长且平行的侧屋，蹲伏在沙丘下面，借以避免强劲西风的侵袭。这种建筑方式与中国人总是把房屋朝向南方，实有异曲同工之妙。因为在某些季候，阳光炙热，暑气迫人，房屋朝南施建，并筑有高大、突出的屋顶，是最为必要的。查考那些遗留下的古旧屋宇，再研究中国传统建筑原理，莫不与天地、鬼神等学说相吻合。这些不成文的条律难以自圆其说，在欧洲只有建造庙宇和礼拜堂时，才会出现相同的"清规戒律"。譬如古代教堂里的牧师席位必须放在礼拜堂的东端，这与经验常识毫无关系，仅是仪制罢了。中国的房屋建筑也很看重仪礼，如果某一件事没有惯例可循，便要向似乎执掌有大自然权力的僧侣们请教咨询。试想，建造一座房屋、一幢庙宇都要依据古制仪礼，那么着手兴建皇朝帝国的京师，该是多么庄重的事情啊！北京实在比任何一国的首都面积都大，这里乃是皇帝驻节之地，而那时的皇帝已经被神化了，被尊为"天子"，替天执行宗教领袖的职责，其权威远高于国王或君主。每年隆冬，皇帝必要亲临天坛向上苍祈祷，求赐丰收。他扮演了人民心目中的精神领袖的形象，他的龙座是神圣不可侵犯的，他的金銮殿也像庙宇似的，朝向正南，整个城市均成为这座庙宇的属地了。

5   Peking's most important feature was the great Chang'an Road, paved with broad slates, which led from the throne hall directly south to that part of the city where the Temple of Heaven and the Temple of Agriculture lay.

City wall and canal surrounding Peking.

6   In its plan Peking is very **reminiscent**[24] of Babylon as described by Herodotus (484 B.C.–425 B.C.) about 450 B.C., another city which had been equally important as capital, trade centre and temple. According to Herodotus, it was a large, regular, walled town built round the Processional Road leading from the palace to the main temple. Recent excavations, however, have proved that his description was somewhat **idealized**[25]. But in Peking the ideal city became reality. Flocks of sheep **graze**[26] peacefully along the **canals**[27] which, **moat**[28] like, surround the city. Within, Peking appears as one square town placed inside another until the holy of holies is reached—"the Forbidden City", where the emperor resided, and where no **intruder**[29] could force his way. In contrast to the colourless residential sections with their grey walls and grey roofs, "the Forbidden City" glitters with red plastered walls, many-coloured woodwork and roofs of glazed **ochre tiles**[30]—ritual again, only the emperor's buildings could have yellow roofs. While it was called by Europeans the Forbidden City, its Chinese name was the Purple City. This name, T'zu Chin Ch'eng, has nothing to do with the colour of its **crenellated**[31] walls but is purely symbolical. It is an **allusion**[32] to the "purple polar star", centre of the celestial world, as the Imperial Palace was the centre round which the terrestrial world **gravitated**[33]. Thus both **Confucius and Laotse**[34] speak of the emperor's position.

　　长安街是北京城最大的特征，街上铺着宽阔的石板，自金銮殿向南直达城的南端，即天坛和先农坛所在地。

环绕着北京的城墙与护城河

　　北京城的都市规划使人联想起约公元前 450 年时希腊历史学家希罗多德（Herodotus，公元前 484 年—公元前 425 年）述及过同样重要的古都——巴比伦，该城是贸易和宗教中心。根据希罗多德的描述，巴比伦城墙巨大而规整，阅兵大街（Processional Road）从宫殿直达庙宇；但最近的考古挖掘证明他的描述有些理想化了，该城并非像他所述的那样壮伟。可是北京却早已把这个理想变为了现实。护城河两岸，成群的牛羊在那儿娴静地吃着青草。北京城是一圈又一圈的城墙所围合起来的，直到最神圣的紫禁城，这里是皇帝的禁宫，皇帝起居所在之处，入侵者是无法逾越的。住宅区苍白黯淡的灰墙、灰瓦顶与紫禁城的红墙、五彩缤纷的柱梁和金黄耀目的琉璃瓦相比，根本不可同日而语。根据礼制，只有皇帝的宫殿才准用黄色屋顶，所谓紫禁城的"紫"字，实际上与颜色无关，只不过暗示着紫极星，表示位居世界中心；皇城位居正中，世界在引力作用下

The Purple City is built symmetrically about the great north-south axis with large reception halls and courts containing apartments for the emperor's **concubines**[35] and **eunuchs**[36] and an enormous court staff. Outside its walls lies the imperial city, also surrounded by walls and also a court city. But within its domain is a special precinct—surrounded by walls of its own—called the Sea Palaces, a fantastic park containing three artificial lakes, with artificial mountains and grottoes, temples, pavilions and houses, where the emperor could live like a philosopher amid natural surroundings—natural surroundings designed and made by man!

The great processional road looking toward the entrance to the Forbidden City.

The White Dagoba crowning a little artificial mountain at the northernmost end of the Sea Palaces.

围绕着皇城。儒家和道家也都尊重皇帝的地位。紫禁城内的建筑完全沿着南北轴线对称而建，紫禁城内建有巨大的厅堂和院落，包括寝宫、后妃们居住的房舍及庭院、太监执事们的住屋。紫禁城以外就是皇城，也有城墙包围着，是一个庭院式的城，在这片区域内，附设着一处游乐地区——另有城墙加以保护——叫作"北海"。这是一个奇妙的花园，包括了三个人工湖泊、人造的土山和假山、庙宇、长廊和房舍。有时皇帝也会像哲人一样到此亲近自然。这里的自然景观都是人工建造起来的。

通向紫禁城入口的宽阔大道

北海最北端人造小沙丘上的白塔

The lakes in the gardens of the Sea Palaces, seen from an artificial mountain north of the Forbidden City. On the left is the little mountain with the White Dagoba.

7 The whole of Peking has been fitted into [37] nature by being given great symmetry. The city itself is an image of nature's undeviating regularity as the astronomer knows it. But the Sea Palaces also represent a special interpretation of nature, the interpretation of the painter and poet which is not dependent on anything so simple as rules. The moment you set foot inside the walls of the Sea Palaces you feel as though you have been transported to a fabulous [38] place far out in the country. From the West Mountains, which from Peking are seen as a blue silhouette [39], water has been brought through canals to the city for the moats and the large artificial lakes. The soil which was removed to make the northernmost lake has been formed into a pleasant little mountain which is crowned by a bottle shaped pagoda [40], the White Dagoba as Europeans call it, a monument for a Buddhistic [41] relic. A bridge leads to it. From the pagoda there is an excellent view over the city, and up here even the symmetrical Forbidden City appears irregular because it is seen from a corner [42]. The great city of a million souls seems to be one huge park for in the glimmer of the sun's rays the view is dominated by the many green trees which overshadow the small, grey houses. Only the gateway towers and the high roofs of the Forbidden City stand out clearly. Up here on the mountain the sun is baking and throws back glaring reflections from the mirror of the lakes. Under leafy trees winding paths lead down its slope [43]. At one spot you find an open wooden

从紫禁城北侧人工山丘上远眺北海风光，左侧是小山和白塔。

可以说整个北京以对称原则巧妙地被安排在大自然的怀抱之中，城市宛如自然一般规整，遨游在太空的宇航员一目了然。而北海又代表了另一种对自然的诠释，一种艺术家和诗人的诠释，这一类型的诠释不是简单地依赖于普通规范。当你踏进北海的入口，你会感到已经到了人间仙境，远离尘世。从北京远眺，西山呈现出一片蓝色的侧影，此处是北京的水源地，护城河与人工湖泊所需的水，都自该水源地经河渠流入城内。为了修建最北侧的湖泊，挖出了大量的泥土，人们利用这些废土筑起了一座可爱的小山，山上还有一座瓶状的宝塔，欧洲人称之为"白塔"，是供奉佛教圣物的建筑物，有桥连通陆地与琼岛。在塔上远眺，全城尽收眼底。因为是从另一个角度看过去的，所以紫禁城内的对称建筑似乎也变成了不规则的状态。拥有百万人口的大城市看起来好似一个大花园，在阳光的普照下，矮小的灰色房屋均隐藏在茂密的绿树中，更添情趣。只有那高耸的城楼和紫禁城内的宫殿，仰着头傲视一切。山上，阳光照耀，湖面似镜，再加上由湖面反射过来的阳光，益增骄阳的威力。树下的山坡上环绕着蜿蜒曲折的小径。你会发现一座开放式的木

gallery through which winged insects fly. It has a tiled floor and in one corner there is an opening with a stairway that seems to descend right into the heart of the mountain. But after a turn the dark, cool grotto ends in a circular wooden pavilion resting on the side of the mountain with a broad view over the lake to the far distant shore. Groping your way further down the slope you reach a little flagged **terrace** [44] where you discover a stone turtle with a tall stone tablet standing on its back and, further still, you come to a long, curved wooden gallery edging the lake. Low skiffs sail among the lotus over to the other shore which is a **legendary** [45] world of tea pavilions, strange temples, bridges and galleries, **weird** [46] rock gardens, and all the rest that a Chinese imagination can invent, executed with great **elaborateness** [47]. It is that Chinese culture we know so well from oriental **porcelain** [48], **embroideries** [49] and painting, and which has inspired Europeans to bring forth a thousand chinoiseries until at last it has come to represent to us the very **quintessence** [50] of Chinese being. How doubly surprising it is, then, when we come out of the **enchanted** [51] world of the Sea Palaces into the everyday greyness of Peking to discover that it is classical and regular, so completely lacking in all that which we mean by chinoiserie!

One of the larger dwellings of Peking, consisting of a number of houses and courtyards.
It is exactly orientated in relation to a north-south axis. But the entrance (at the bottom of the drawing) is placed to one side of the axis. This is a fixed principle: evil spirits must not be able to rush straight into the house.
Note that the northern gates of Peking are not placed opposite the southern gates.

游廊，昆虫上下飞舞。游廊的地面上铺着地砖，拐弯出口处安置着楼梯，由此而下可直达山中。转了一个弯，走过了一段又暗又凉的山洞，出得洞来，便是依山修筑的一条圆形木制长廊。站在廊上，视界开阔，全湖尽收眼底，还可遥望对岸景色。再向坡下走一段，便进入一座用石板铺砌成的阳台，阳台上有一只石龟，龟背上驮着一块大石碑。再前行抵达湖滨，有一座木造的蜿蜒走廊，供游人休息并可登小艇游湖。湖中遍植荷花，小船在荷丛的芳香气息中穿行，不知不觉抵达彼岸。在那里又进入另一个传奇世界，岸上有茶棚，有陌生的寺庙，有石桥亭台，有奇石园……凡中国人所能想到的造景元素，都极尽所能地汇集到了一起。我们通过东方的瓷器、刺绣和绘画，了解了中国文化，创造出具有中国艺术风格的制品，"代表"了中国文化的精髓。当我们走出令人痴迷的北海，踏入北京平凡的灰色民居时，才惊讶地发现我们所定义的"中国艺术风格"完全缺失了经典与规律的特质。

北京城内的大宅院，有许多房间和院落，依照东方习俗坐北朝南。

图中住宅最下端的大门，没有设置在南北轴线上，这是一条固定的原则——阻止妖魔直接进入庭堂。

请注意：在北京，凡是朝北的门不会和朝南的门相对而设。

8   The great main highways are very broad avenues with a roadway in the middle for all through traffic. On each side of it there is an equally broad **thoroughfare** [52] which partly serves as what we call the sidewalk, partly is used for more local traffic, partly is an area for all sorts of outdoor work. It is an excellent arrangement. Shops and work rooms are open booths which almost **merge into** [53] the street where the Chinese display their wares, shoe horses, hold their pig market, let their tired camels sink heavily down to rest, fetch up the water from the public wells, are shaved and trimmed by the street barber, and carry on a hundred and one other things. Originally the middle roadway was reserved for the emperor's officials and **the mandarins** [54]. Everything, indeed, had its purpose in relation to the imperial residence. For instance, the reason why houses could be only one storey high was so that no one could see over the walls of the Forbidden City.

9   From the broad highways run the narrow residential streets where there are no shops. There may be houses which reflect the whole plan of Peking with everything carefully orientated in relation to a north-south axis and with several courtyards, one behind another. There may also be unsymmetrical Chinese gardens behind symmetrical groups of buildings. But seen from the street there is hardly any difference to be noted between a large, costly **mansion** [55] and a cluster of coolie hovels. One sees only grey walls without windows, with here and there an entrance. Just as no one may look over the walls of the imperial city, no one may compete with the magnificence of the imperial palace. In the small, narrow streets there are no shops. But they are visited by many dealers carrying their wares. Each one calls attention to himself by means of some small instrument, a flute, a long **tuning-fork** [56], small brass bowls which are sounded against each other (something like castanets); all make delicate sounds that are heard behind the walls in faint tones which announce to the listener who it is that is passing.

10   It is an existence and a world very different from ours. And yet it is of great interest to us, not as a curiosity but as the most fully developed example of a special city type which arises when a ruler makes himself a **high-priest** [57] or a **deity** [58] for his people. Of itself a ritual develops which forms the city. When the French monarch became absolute a system of rules and regulations evolved governing his conduct and the conduct of those surrounding him, it was logical that he should live in a city which was just as systematically built up around the king.

北京的主要道路是很宽阔的大街，中央部分用来行驶车辆；两侧还有等宽的道路，作为人行道，兼作慢速行驶车辆之用，还可以供各种户外作业使用。这是一种非常优秀的道路设置。两旁的商铺或作坊属于几乎已融入大街的开放摊位，人们利用门前区域陈列货品或给骡马钉掌，也可买卖家禽家畜。长途跋涉后的骆驼也能在此躺下休息一会儿。人们在附近的水井汲水供牲口饮用。还有流动的剃头摊子给路人理发，此外还有很多交易是在这路边的场地上进行的。道路的中央部分，原本是保留专供官吏或贵族们使用的，这里的一切都与皇宫有着密切的关系。譬如普通民宅一律是平房，因为一旦建造了高楼大厦，紫禁城内的秘密就一览无遗了。

从主要道路进入住宅区域，只有狭窄的胡同小巷，见不到商号了。有些宅院方方面面都精心符合南北朝向，反映北京城的整体规划，附带有一进套一进的几进院子。有时在这些互相对称的房屋后面，却有一个不对称的中式花园。从街上看来，深宅大院与平民住屋没有太显著的分别，一律是灰色的围墙，墙上无窗，只是偶尔开有一扇门罢了。因为看不到皇宫内的情形，根本无法将外面的宅院与金碧辉煌的皇宫相比。胡同小巷内，没有店铺，却有不少商贩挑着货物走街串巷。他们都带着一件能发出声响的器具来表明自己的身份，一支笛子、一把音叉、一面铜锣，各能发出独特的声音，使深居墙内的人闻其声便确知是何种商贩在此经过。

这些风土人情和西方世界迥异，但对西方人士颇具吸引力，不仅仅是好奇；西方人对在兼有宗教领袖身份或被奉若神明的最高统治者的主持下施建的最为健全的特殊城市类型产生了浓厚兴趣。当法国君主成为绝对的专制君主，有权来改善生活环境后，他也必然要建造一个以他为中心的城市。

Residential streets "Hutungs" in Peking. The street is not paved, only a dusty grey road between grey houses with here and there a large green tree seen above the walls. Lean black pigs gallop through the streets and in the summer well-built, stark-naked children play in them with no unnecessary noise while the tradesmen pass through the streets producing their delicate, elfin-like music.

北京住宅区的小胡同没有铺设路面，而是土路。在灰色的房屋之间，时常有高大的绿色树冠伸出墙外。

黑瘦的猪也在胡同内东游西窜。夏天身体健壮的孩子们光着身子在一起玩耍。

商贩们经过时又会弄出音乐般悦耳的声响，引起居民注意。

（文献来源：斯坦·埃勒·拉斯穆森.城镇与建筑.韩煜，译.天津：天津大学出版社，2013）

词汇 | Vocabulary

[1] illuminative 照亮的，有启发性的

[2] metropolis 大都市

[3] storeyed 有……层楼的 / one-storeyed 一层楼的

[4] coexistent 共存的

[5] residential 住宅区，居民区

[6] grandeur 壮丽

[7] conviction 深信的观点

[8] orientation 定位，方位

[9] squat down 蹲下

[10] dune 沙丘

[11] fierce 强烈的，激烈的

[12] abracadabra 胡言乱语，此处指天地鬼神等传说

[13] imperceptibly 极微地，察觉不到地

[14] imperative 极重要的

[15] ritual 仪制，典礼

[16] priest 祭司

[17] regent 摄政王

[18] sovereign 君主

[19] pontiff 教皇

[20] prerogative 权力

[21] sacrifice 供奉，献祭

[22] omnipotent 万能的

[23] throne （君王的）宝座

[24] reminiscent 怀旧的

[25] idealized 理想化的

[26] graze 吃草

[27] canal 运河

[28] moat 护城河

[29] intruder 入侵者

[30] ochre tile 赭色的瓷砖

[31] crenellated 锯齿状的

[32] allusion 影射

[33] gravitate 受重力影响而运动

[34] Confucius and Laotse 孔子与老子，儒家与道家

[35] concubine 妃嫔

[36] eunuch 太监

[37] fit into 刚好放入，符合

[38] fabulous 极好的

[39] silhouette 侧影，轮廓

[40] pagoda 佛塔

[41] Buddhistic 佛教的

[42] corner 角，角落

[43] slope 斜坡

[44] terrace 阳台

[45] legendary 非常有名的

[46] weird 奇怪的

[47] elaborateness 尽心竭力

[48] porcelain 瓷器

[49] embroidery 刺绣

[50] quintessence 精华

[51] enchanted 被施魔法的，令人痴迷的

[52] thoroughfare 大道

[53] merge into 并入，融入

[54] mandarin 官员

[55] mansion 大厦

[56] tuning-fork 音叉，一种乐器，此处指货郎所用的梆子、木鱼、拨浪鼓等

[57] high-priest 大祭司

[58] deity 神

练习与思考 | Comprehension Exercise

熟读课文，了解西方学者对北京城的认知，掌握预备知识中的中国城市建设史及文化遗产术语。

课后延伸 | Reading Material

古今北京 | The Traditional and Modern Beijing

The Forbidden City

1   Built as the imperial palace of the Ming Dynasty, the Forbidden City has served as the residence of the emperor and his household as well as the ceremonial and political hub of the Chinese government. Organized to establish levels of hierarchy and ceremony, the extensive compound consists of nearly one thousand buildings. Positioned with strict hierarchies along a dramatic central axis, the buildings are governed by a relentless symmetry that repeats and maps the geometric purity of the balanced plan. The resulting composition is a ceremonial complex that is heroic in both scale and effect. The organization allows for the hierarchies of social and political interaction to be exquisitely orchestrated.

2   The Forbidden City is defined by its axial hierarchy. Surrounded by fortress walls, the city is organized along a central axis running from the Meridian Gate, passing through a number of halls, and terminating in the Imperial Garden. The axis of the Forbidden City was designed for the emperor. Only he was allowed to travel along the axis, thereby making the organizational aspect of the complex more powerful through its combined spatial, political, and religious agendas. The form prevalent in the architecture was similarly reinforced through its use, governing the structure of ceremonies and events.

紫禁城

紫禁城始建于明代，是皇帝的住所和皇室家族的居住地，也是中国的礼教中心和政治中心。庞大的紫禁城有近千个建筑，组成不同的院落空间，展示了不同的等级制度和礼教制度。紫禁城中建筑的布局沿中轴线严格对称，体现出森严的等级秩序。反复利用纯粹的几何图形进行平面布局，形成了具有平衡感的总体平面，由此最终形成了紫禁城这一处具有强烈仪式感的皇家建筑群。这种仪式感又使得建筑群更显规模庞大、恢宏壮丽，对后世也产生了深远的影响。同时，中轴对称的组织形式也实现了不同社会等级和政治群体的毗邻而居与和谐共存。

紫禁城是依据轴线等级建立起来的。整个紫禁城被宫墙环绕，沿着中轴线展开空间组织，中轴线自午门起始，贯穿多个宫殿，最后止于御花园。紫禁城的轴线是为皇帝一人服务的。只有皇帝可以在中轴线上行进，因此通过轴线与空间形制、政治体制和精神信仰的结合，紫禁城皇家建筑群的组织性更强。通过中轴线的使用，建筑的形制等级也得以增强，并统领了仪典礼制的等级结构。

Chinese Courtyard House

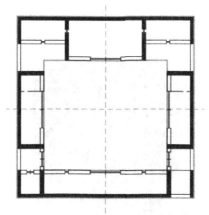

1 The courtyard, found in numerous countries and cultures, is one of the most ubiquitous architectural elements. At its most basic level, it is a way to bring the exterior into the center of the house. It provides the heart of the house through a space that relates to many of the functions and rooms. It is a type that has crossed centuries and known few boundaries in terms of materiality or construction. It is difficult to imagine an urban context that does not have courtyards. The courtyard allows the building to become a series of thin layers that permit light and air to enter into the most interior spaces.

2 The Chinese Courtyard House, or siheyuan, focuses upon the geometric purity of the inner courtyard as a square and establishes a distinct object-like quality to the surrounding pavilions, each laid out along a north-south and east-west axis. Thus, unlike the subtractive and additive relationships of the European and Middle Eastern traditions, the courtyard is formed through the aggregation of units around a space. Made of repetitive and unitized pieces of wood and masonry, the additive, elemental, assembly-based construction system allows for distinct compositional segmentation that carries through from part to whole. Each house around the courtyard maintains independence of ownership often belonging to a different member of the family. The courtyard itself is for privacy and contemplation as opposed to community, often with multiple courts receding into the site, offering more privacy.

四合院

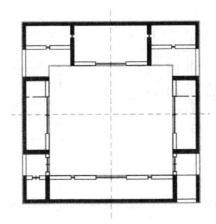

在众多国家和文化中，庭院都是最普遍的建筑元素之一。究其本质，庭院是一种将外部空间引入建筑内部的方式。庭院是一处与众多功能及众多房间都存在联系的空间，因此庭院可以成为建筑的核心空间。庭院这一类型已经跨越了几个世纪，但其物质特性和建造结构却鲜为人知。很难想象在城市文脉中如果没有庭院会是什么样子。庭院能将光和空气引入建筑内部，也能使单薄的建筑层次形成有秩序的空间序列。

中国的四合院十分注重内部方形庭院几何形式的纯粹性，四合院中的建筑一般沿着南北轴和东西轴布局，而方形庭院营造出的空间品质与周围环绕的建筑明显不同。因此，不同于欧洲和中东地区通过加减关系创造庭院，中国四合院是由围绕着庭院空间的建筑单体聚合而成的。在其施工体系中，以重复的组合式木质构件和砖石砌块建造，辅以反映自然力量的装配式构件系统，无论在局部还是整体上都实现了明显的构图分割。院子里的每间房屋均能保持其所有权的独立性，通常不同的房屋归属于不同的家庭成员。院子本身是私密的，适合冥想沉思，这一点与公共社区截然相反，四合院中常常还有多重院落层层后退，以提供更强的隐私空间。

Tongzhou Gatehouse

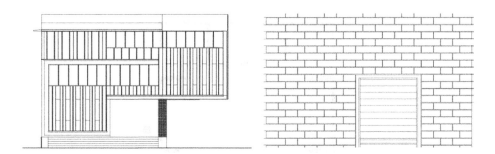

In the design of Tongzhou Gatehouse, Office dA used local gray bricks in ways that were expressive and novel. They were materials suited for both running and stacked bonds and the removal of certain bricks could create textual patterns and produce formally dynamic and dematerialized walls. These operations created a broad variety in the otherwise blank surfaces. The material emphasis was translated to a detailing system of architecture that allowed for the celebration of the beauty of the brick, as a structural material and an ornamental element simultaneously.

通州艺术中心门房设计

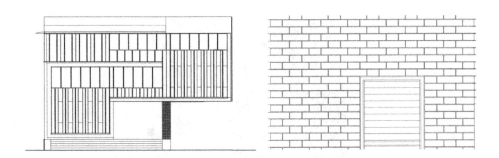

　　在通州艺术中心门房设计中，dA 建筑师事务所创新性地使用了一种当地的灰砖，既新奇又具表现力。这种材料能够适应顺砖砌法和对缝砌法，还能通过一些砖块的移除创造一种肌理图案，并产生动态的、非实体的墙壁形式。这些设计手法在原本单调的建筑表面上创建出多种多样的形式。在该建筑中，灰砖既是结构材料，又是装饰元素，建筑材料所强调的重点转变为建筑的细部体系，彰显了砖材之美。

CCTV Building

The CCTV Building employs a diagonal structure grid in the exterior that creates a proportional pattern that is changing. To display the varistructured forces of the irregular building form, the skin pattern has been necessarily branched and encrypted. Talking into account the combined requirements of the overall form of the building and the functional practicality of its physical features, the pattern in the building skin clearly expresses the intrinsic natural forces. This pattern is regulated by the simplicity of proportionality as the response variable, and brings about an effect of both functionality and decorativeness.

中央电视台总部大楼

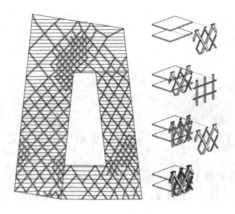

　　中央电视台总部大楼采用了一种外部可见的对角式结构网格，形成了一种变化的比例模式。为了使不规则的建筑形式展现出可变结构的力量，建筑表面的图案进行了必要的分支加密处理。考虑到建筑的整体形式和物理特征的功能实用性这二者的综合要求，建筑表面的图案模式鲜明地表达出各种内在的自然作用力。这种模式以比例作为响应变量进行调节，简单又直接。最终所创造的表面图案既是功能性的反映，又具有装饰效果。

The Bird's Nest Stadium

1   The Bird's Nest Stadium, built for the 2008 Beijing Olympics, is visually dominated by a series of seemingly random steel members that wrap around the entirety of the stadium, thus giving the complex its name. These steel beams form an exoskeleton that not only acts as an appliqué or ornament, but also performs in a structural manner, helping to support various parts of the perimeter of the stadium. It is essentially an applied aesthetic solution that is governed by compositional sensibilities; however, it substantiates itself further by solving problems of statics and structure.

2   The Bird's Nest adopts the plan typology of the Roman Coliseum. Designed to host the track and field events, the elliptical running track establishes the primary inner form reiterated through the surrounding ring of seating. The adoption of a functional and historically predetermined planning typology fits well with the agenda and methodology of Herzog and de Meuron. Their architecture is defined through the development of skin and the collaborative ingenuity of material, tectonic, surface, and pattern to produce identity on a traditional and simplistic form. In the case of the Bird's Nest Stadium, densely lapping and intersecting bands non-orthogonally intersect and aggregate to define a porous and homogenous yet dynamically individuated shell.

鸟巢体育场

　　鸟巢体育场是为了 2008 年北京奥运会而设计建造的，体育场外表主要由一系列看似随意的钢结构缠绕着，而体育场也因此得名。这些钢铁梁架形成了建筑的外部骨架，它们不仅作为附加的建筑装饰或饰物，也具有一定的结构性作用，帮助支撑体育场边缘的各个部分。从实质上说，它是一种实用的美学解决方案，由建筑创作的感知力所主导，然而，它也证实了钢铁梁架本身确实具有进一步解决静力学问题和结构问题的能力。

　　鸟巢体育场参照了古罗马斗兽场的平面设计风格。鸟巢体育场主要承担田径赛事，椭圆形的跑道与椭圆形的看台结合，共同建立起体育场主要的内部形态。古罗马斗兽场这种在历史上早已应用、在功能上业已成熟的平面类型，能够很好地适应赫尔佐格和德·梅隆事务所的设计思路和方法。他们的建筑作品通过对建筑外表面的研发，再结合对材料、结构、表面及图案模式的改良和创新，形成了既符合传统又简洁独特的个性。在鸟巢体育场这个案例中，结构网架通过高密度的搭接和非正交的相交聚合在一起，形成了一种多孔、同质、均匀、富有动态的个性化外壳造型。

　　（文献来源：盖尔·彼得·博登，布莱恩·代尔福德·安德鲁斯.建筑原理：形式材料的基本原则.杨慧，译.天津：天津大学出版社，2021）

第 2 课　城市与区域发展
Lesson 2　Urban and Regional Development

导读 | Introduction

《伦敦规划 2021》 | *London Plan 2021*

英国是第一次工业革命的发源地，18 世纪的工业革命迅速推动了城市化的发展。伦敦在城市扩张的进程中，颁布了一系列有关城市卫生和工人住房的法律法规，这些文件孕育了第一部关于城乡规划的法律——《住房、城镇规划等法》（1909），这是现代城市规划确立的标志，标志着城市土地开发控制的实践和理论研究成为政府职能，标志着城市规划纳入政府管理。

2021 年 3 月《伦敦规划 2021》（*London Plan 2021*）向社会公布，内容涵盖经济、环境、交通与社会，代表了伦敦的总体战略计划，是未来 20~25 年伦敦城市发展的综合框架。伦敦是欧洲最大的城市，其城市规划一直走在世界前列，引领着全球城市的规划理念与方法。《伦敦规划 2021》是伦敦市的最新探索，其确定了城市中的增长潜力区、伦敦市中心的活力计划、公交基础设施的连通计划、可负担的新住房计划，明确了新建筑的"零碳"建造标准以及标准开发的"循环经济"原则，对解决住房危机（the housing crisis）、应对气候紧急状态（the climate emergency）、建设绿色健康城市（a greener and healthier city）、维护城市特色与文化遗产（character and heritage）、疫情后的城市恢复（recovery from COVID-19）都提出了详细的计划。

预备知识 | Preliminary Knowledge

《伦敦规划 2021》简介 | Introducing *London Plan 2021*

《伦敦规划 2021》（*London Plan 2021*），即《伦敦空间发展战略》（*Spatial Development Strategy, SDS*）。根据伦敦政府的相关立法，伦敦市市长应发布空间发展战略并进行审查。《伦敦规划 2021》是伦敦的总体战略规划，是伦敦未来 20~25 年的综合的经济、环境、交通和社会框架。

《伦敦规划 2021》的战略重要性如下。

· 促进大伦敦地区的经济发展和财富创造；

· 促进大伦敦地区的社会发展；

· 促进大伦敦地区环境的改善。

《伦敦规划 2021》需考虑以下因素。

· 符合人人机会平等原则；

· 减少健康不平等，提高伦敦人民健康水平；

· 实现英国可持续发展；

· 应对气候变化及其影响；

· 促进和鼓励对泰晤士河的利用，特别是增加客运和货运的诉求；

· 整合资源以增强实施性。

《伦敦规划 2021》是伦敦市市长和伦敦 32 个自治市镇、伦敦金融城公司（the City of London Corporation）、市长发展公司（the Mayoral Development Corporations, MDCs）、当局指定的社区议事中心的共同责任。当《伦敦规划 2021》提到"自治市镇"（boroughs）时，指的是伦敦的 32 个自治市镇、伦敦金融城公司和市长发展公司。

《伦敦规划 2021》的期限为 2019 年至 2041 年。

《伦敦规划 2021》的性质如下。

《伦敦规划 2021》是一项全新的规划，在此之前于 2004 年、2011 年编制了两次。本次《伦敦规划 2021》与之前的规划不同，它更加雄心勃勃，目标更加明确。美好增长（Good Growth）——具有社会与经济包容性以及环境可持续性的增长——是《伦敦规划 2021》的基础。

《伦敦规划 2021》的综合影响评估如下。

《伦敦规划 2021》评估的关键部分是进行全面的综合影响评估（Integrated Impact Assessment, IIA）和栖息地法规评估（Habitats Regulations Assessment, HRA）。

其中，综合影响评估（IIA）指通过一个综合过程对规划及其拟议政策进行全面评估，包括：

· 战略环境评估；

· 可持续性评估；

· 平等影响评估；

· 健康影响评估；

· 社区安全影响评估。

栖息地法规评估（HRA）将评估《伦敦规划 2021》中任何可能对欧洲栖息地造成重大影响的方面。

《伦敦规划 2021》的结构如下。

· 第一章　城市愿景

· 第二章　伦敦整体空间发展布局

· 第三章　设计

· 第四章　住房

· 第五章　社会基础设施

· 第六章　经济

· 第七章　遗产与文化

· 第八章　绿色基础设施和自然环境

· 第九章　可持续基础设施

· 第十章　交通

· 第十一章　提供资金

· 第十二章　监测

课文讲解 | Text Explanation

Good Growth · Planning London's Future

1 Good Growth [1]—growth that is socially and economically **inclusive** [2] and environmentally sustainable—**underpins** [3] the whole of the *London Plan 2021* and each policy. It is the way in which sustainable development in London is to be achieved.

2 London's growth and development is shaped by the decisions that are made every day by planners, planning applicants, **decision-makers** [4] and Londoners across the city. Every individual decision to provide affordable housing helps to make the housing market fairer. Every decision to make a new development car-free helps Londoners to depend less on cars and to live healthier lives. Every decision to build or expand a school improves the prospects of the next generation of Londoners.

美好增长 · 伦敦的未来

　　美好增长（Good Growth，GG），是指具有社会与经济包容性以及环境可持续性的增长，是整个《伦敦规划 2021》以及各项政策的基础，是伦敦实现可持续发展的方式。

　　伦敦的进步和发展，是由每一位规划师、规划申请人、决策者和全体伦敦市民的日常决定所成就的。每一个提供经济适用住宅的决定都有助于使住房市场更加公平，每一个开发项目无车化的决定都有助于减少伦敦对汽车的依赖，每一个新建或扩建学校的决定都会改善下一代伦敦人的未来。

Each week, hundreds of these individual decisions contribute to progress across London, shaping places and improving lives. Over the course of years, they can transform the whole city for the benefit of Londoners now and in the future.

3 The *London Plan 2021* provides the strategic framework within which all these decisions are made. It guides boroughs' development plans to ensure that they are working towards a shared vision for London, and it establishes policies that allow everyone involved in new developments to know what is expected from them.

4 This *London Plan 2021* will help realize the Mayor's vision of creating a city for all Londoners, where no one is left behind.

Planning for Good Growth

5 The *London Plan 2021* covers the full range of planning issues, from the number of homes London needs to the design of its streets, and it is important that each policy is clear and implementable.

6 London's global economy is the envy of other world cities and with good reason—it is the engine of the national economy and will sustain the level of population growth expected in London over the coming years. But to plan a city that works for all Londoners, as the population grows towards 10.8 million by 2041, it will be important to think about what the purpose of economic growth actually is.

7 A failure to consider this fundamental question has led to some of the most serious challenges London faces today. The growth in population and jobs has not been matched by the growth in the number and type of homes London needs, driving up rents and house prices to levels that have priced many Londoners out of the market. A focus on large multinational businesses in the centre of London has not been matched by economic development in other parts of the city. A failure to consider the wider implications of London's growth has increased car dependency, leading to low levels of physical activity, significant congestion, poor air quality and other environmental problems.

每周都有数百个这样的决定，为整个伦敦的发展作出贡献，塑造场所并改善生活。多年以后，它们就可以改变整个城市，造福现在和未来的伦敦人。

《伦敦规划2021》为所有这些决定的制定提供了战略框架。它指导各行政区的开发建设规划，以确保所有项目服务于伦敦的共同愿景。同时，它制定政策，让新项目的参与者了解应达到的目标。

《伦敦规划2021》将有助于实现市长愿景，即为所有伦敦人创造一个公平共享、人人可期的城市。

为"美好增长"而规划

《伦敦规划2021》涵盖了从伦敦住房需求到街道设计等全方位的规划议题，每一项政策都明确、可实施。

伦敦是英国经济的重要引擎，其经济的全球化让其他世界级城市羡慕不已。伦敦将维持其人口增长水平，预计人口将在2041年增至1 080万。要规划一个适合所有伦敦人的城市，思考经济增长的真正目的就显得至关重要。

如今伦敦也面临着一些挑战，如人口和工作岗位的增长与伦敦所需住房数量和类型的不匹配，租金和房价上涨，城市中心集中发展大型跨国企业而其他区域的经济发展却没有跟上，对汽车的依赖性过高导致低运量、交通拥堵严重、空气质量下降以及其他环境问题。

8　　This *London Plan* takes a new **approach** [5]. It plans for growth on the basis of its potential to improve the health and quality of life of all Londoners, to reduce **inequalities** [6] and to make the city a better place to live, work and visit. It uses the opportunities of a rapidly-growing city to plan for a better future, using each planning decision to improve London, transforming the city over time. It plans not just for growth, but for Good Growth – sustainable growth that works for everyone using London's strengths to overcome its weaknesses.

9　　A city that is planned well can improve as it grows. Planning for the right number of homes and higher levels of affordable housing will take advantage of London's growth to re-balance the housing market. Planning for **mixed-use developments** [7] in all parts of London will spread the success of London's economy and create stronger communities where everyone feels welcome.

10　　Planning new developments to reduce car dependency will improve Londoners' health and make the city a better place to live. Planning for a 'smarter' city, with world-class digital connectivity will enable secure data to be better used to improve the lives of Londoners.

11　　To ensure that London's growth is Good Growth, each of the policy areas in this Plan is informed by six Good Growth objectives:

　• GG1　Building strong and inclusive communities

　• GG2　Making the best use of land

　• GG3　Creating a healthy city

　• GG4　Delivering the homes Londoners need

　• GG5　Growing a good economy

　• GG6　Increasing efficiency and resilience

12　　Planners, planning applicants and decision-makers should consider how their actions are helping to deliver these objectives as they work to develop and improve London. By doing so, they will keep London's development on track, ensuring that the growth of the city benefits all Londoners.

《伦敦规划 2021》采用了一种新的规划方法。它依据"改善所有伦敦人的健康和生活质量"的目标来规划城市的增长，减少不平等，并使城市成为一个更美好的生活、工作和旅游场所。《伦敦规划 2021》将利用城市所拥有的机遇来规划更美好的未来，利用每一项规划决策来发展伦敦。这份规划不仅仅是为了增长，而是为了"美好增长"，即利用伦敦的优势来实现适合所有人的可持续增长。

一座经过周密规划的城市能在其发展的过程中不断得到完善。规划适当数量的住房和更高水平的经济适用住宅将重新平衡住房市场。所有混合用途开发项目（mixed-use developments）都将推广伦敦经济的成功，并创造包容每一个人的更强大的社区。

新开发项目将减少对汽车的依赖，改善伦敦人的健康状况，使城市更适合生活。规划一个拥有世界级数字连接的"更智慧"的城市将使安全数据更好地被用于改善伦敦人的生活。

为了确保伦敦的发展是"美好增长"，本规划文件中的每一个政策都基于六个"美好增长"目标：

- GG1 建设强大和包容的社区
- GG2 充分利用土地
- GG3 创建健康城市
- GG4 提供伦敦人需要的住房
- GG5 发展良性经济
- GG6 提高效率和韧性

规划师、规划申请人和决策者在努力发展和改善伦敦时应考虑他们的行动如何帮助实现这些目标。这将使伦敦的发展保持在正确的方向上，确保城市的发展惠及所有伦敦人。

GG1 Building strong and inclusive communities

13 London is made up of diverse communities. Its neighbourhoods, schools, workplaces, parks, community centres and all the other times and places Londoners come together give the city its cultural character and create its future. Planning for Good Growth means planning with these communities – both existing and new – making new connections and eroding inequalities.

14 London is one of the most diverse cities in the world, a place where everyone is welcome. 40 per cent of Londoners were born outside of the UK, and over 300 languages are spoken here. 40 per cent of Londoners are from Black, Asian and Minority Ethnic (BAME) backgrounds, and the city is home to a million EU citizens, 1.2 million disabled people, and up to 900,000 people who identify as LGBT+. Over a fifth of London's population is under 16, but over the coming decades the number of Londoners aged 65 or over is projected to increase by 90 per cent. This diversity is essential to the success of London's communities. To maintain this London must remain open, inclusive and allow everyone to share in and contribute towards the city's success.

15 London is one of the richest cities in the world, but it is also home to some of the poorest communities in the country, with wealth unevenly distributed across the population and through different parts of the city. It is home to an aging population, with more and more people facing the barriers that already prevent many from participating fully in their communities. Traffic dominates too many streets across the city, dividing communities and limiting the interactions that take place in neighbourhoods and town centres.

16 Delivering good quality, affordable homes, better public transport connectivity, accessible and welcoming public space, a range of workspaces in accessible locations, building forms that work with local heritage and identity, and social, physical and environmental infrastructure that meets London's diverse needs is essential if London is to maintain and develop strong and inclusive communities.

17 Early engagement with local people leads to better planning proposals, with **Neighbourhood Plans**[8]

GG1 建设强大和包容的社区

伦敦是由形形色色的社区组成的。邻里、学校、工作场所、公园、社区中心以及其他所有伦敦人聚集在一起的时间和地点，都赋予了这座城市文化特征并创造了城市的未来。为"美好增长"而规划，就意味着与这些社区（包括既有和新建社区）一起努力建立新的连接，消除不平等。

伦敦是世界上最多元的城市之一，欢迎所有人。40% 的伦敦人出生在英国以外，有 300 多种语言在这里同时使用。40% 的伦敦人来自黑人、亚裔和少数族裔背景，同时这座城市也是 100 万欧盟公民、120 万残障人士以及多达 90 万的性少数群体（LGBT+）的家。16 岁以下人口超过五分之一，但在未来几十年内 65 岁及以上的人口数量预计增加 90%。人口的多样性对伦敦社区的成功至关重要。为了保持这种多样性，伦敦必须保持开放、包容，让每一个人都能分享这座城市的成功并为之努力。

伦敦是世界上最富有的城市之一，但是由于财富在人群和城市不同区域的分配并不均匀，英国最贫穷的社区也在伦敦。伦敦是老龄人口的家园，有越来越多的人面临出行障碍，不能充分参与社区活动。机动车在街道上占据着主导地位，分隔了社区，限制了邻里和城镇中心的互动。

伦敦要保持并发展强大和包容的社区，提供优质的、经济适用的住房，更好的公共交通网络，便利舒适的公共空间，可到达的工作空间，与当地遗产和特质相适应的建筑形式，以及满足伦敦多元需求的社会、物质和环境基础设施，都是至关重要的。

当地居民的早期参与可以带来更好的规划建议，邻里计划（Neighbourhood Plans）为社区提

providing a particularly good opportunity for communities to shape growth in their areas. Taking advantage of the knowledge and experience of local people will help to shape London's growth, creating a **thriving**[9] city that works better for all Londoners.

The Implementation Strategies of "GG1 Building strong and inclusive communities"

18　Good growth is inclusive growth. To build on the city's tradition of openness, diversity and equality, and help deliver strong and inclusive communities, those involved in planning and development must:

A　encourage early and inclusive engagement with **stakeholders**[10], including local communities, in the development of proposals, policies and area-based strategies;

B　seek to ensure changes to the physical environment to achieve an overall positive contribution to London;

C　provide access to good quality community spaces, services, amenities and infrastructure that accommodate, encourage and strengthen communities, increasing active participation and social integration, and addressing social isolation;

D　seek to ensure that London continues to generate a wide range of economic and other opportunities, and that everyone is able to benefit from these to ensure that London is a fairer, more inclusive and more equal city;

E　ensure that streets and public spaces are consistently planned for people to move around and spend time in comfort and safety, creating places where everyone is welcome, which foster a **sense of belonging**[11], which encourage community **buy-in**[12], and where communities can develop and thrive;

F　promote the crucial role town centres have in the social, civic, cultural and economic lives of Londoners, and plan for places that provide important opportunities for building relationships during the daytime, evening and night time;

G　ensure that new buildings and the spaces they create are designed to reinforce or enhance the identity, legibility, permeability, and inclusivity of neighbourhoods, and are resilient and adaptable to changing community requirements;

H　support and promote the creation of a London where all Londoners, including children and young people, older people, disabled people, and people with young children, as well as people with other

供了一个特别好的塑造本地增长的机会。充分利用本地居民的知识和经验将有助于塑造伦敦的发展，创造一个更好地为所有伦敦人服务的繁荣的城市。

"GG1 建设强大和包容的社区"实施策略

美好增长是包容性的增长。为了发扬这座城市开放、多元和平等的传统，帮助建设强大和包容的社区，规划和开发的相关参与者必须：

A 鼓励包括当地社区在内的利益相关方尽早以包容的方式参与制定方案、政策和地区性战略；

B 力求确保物质环境的改变能为伦敦的整体发展作出积极贡献；

C 提供机会使用优质社区空间、服务、生活设施和基础设施，为社区服务、赋能，提高参与积极性及社会融合程度，并解决社会孤立问题；

D 力求确保伦敦继续创造广泛的经济机遇和其他机会，并确保每个人都能从中受益，以保证伦敦成为一个更公平、更包容、更平等的城市；

E 确保街道和公共空间的规划始终如一，让人们在舒适和安全的环境中活动和消磨时间，创造出人人受欢迎的场所，培养归属感，鼓励社区认同，使社区能够发展和繁荣；

F 促成城镇中心在社会、市民、文化和经济生活中的决定性作用，建设在白天、晚上、夜间都能为建立社会关系提供重要机会的场所；

G 确保新建筑及其营造的空间在设计上能够增强或提高社区的特质、可识别性、渗透性和包容性，并能适应不断变化的社区要求；

H 支持和促进建立一个让所有伦敦人，包括儿童和年轻人、老年人、残障人士、携幼人群以

protected characteristics, can move around with ease and enjoy the opportunities the city provides, creating a welcoming environment that everyone can use confidently, independently, and with choice and dignity, avoiding separation or segregation;

I support and promote the creation of an inclusive London where all Londoners, regardless of their age, disability, gender, gender identity, marital status, religion, race, sexual orientation, social class, or whether they are pregnant or have children, can share in its prosperity, culture and community, minimising the barriers, challenges and inequalities they face.

GG2 Making the best use of land

19 London's population is set to grow from 8.9 million today to around 10.8 million by 2041. As it does so, employment is expected to increase on average by 49,000 jobs each year, reaching 6.9 million over the same period. This rapid growth will bring many opportunities, but it will also lead to increasing and competing pressures on the use of space. To accommodate growth while protecting the Green Belt, and for this growth to happen in a way that improves the lives of existing and new Londoners, this Plan proposes more efficient uses of the city's land.

20 The key to achieving this will be taking a rounded approach to the way neighbourhoods operate, making them work not only more space-efficiently but also better for the people who use them. This will mean creating places of higher density in appropriate locations to get more out of limited land, encouraging a mix of land uses, and co-locating different uses to provide communities with a wider range of services and amenities.

21 High-density, mixed-use places support the clustering effect of businesses known as 'agglomeration' [13], maximising job opportunities. They provide a critical mass of people to support the investment required to build the schools, health services, public transport and other infrastructure that neighbourhoods need to work. They are places where local amenities are within walking and cycling distance, and public transport options are available for longer trips, supporting good health, allowing strong communities to develop, and boosting the success of local businesses.

及其他需要保护的人，都能畅行无阻并享有这座城市机会的伦敦，同时创造一个人人都能自信、独立、有选择、有尊严使用的温馨环境，避免分离和隔离；

笔者支持和促进建立一个包容的伦敦，让所有伦敦人，不论其年龄、健康、性别、性别认同、婚姻状况、宗教、种族、性取向、社会阶层如何，是否怀孕生育，都能分享伦敦的繁荣、文化和社区，最大限度地减少人们面临的障碍、挑战和不平等。

GG2 充分利用土地

到 2041 年，伦敦人口将从目前的 890 万增长到约 1080 万。在此过程中，每年预计平均增加 4.9 万个就业岗位，在 2041 年达到 690 万。这样的快速增长将带来许多机会，但也会导致空间利用方面不断增加的竞争压力。为了适应增长同时保护城市绿带（the Green Belt），并使这种增长以改善现有伦敦人和新伦敦人生活的方式进行，本规划建议更有效地利用这座城市的土地。

实现这一目标的关键是对社区的运作形式的更新，使其不仅能更有效地利用空间，而且能更好地为使用者服务。这意味着在适当的地点创造更高密度的场所以更多地利用有限的土地，鼓励混合利用土地，并将不同的用途集中设置，为社区提供更广泛的服务和设施。

高密度、混合用途的场所可以形成"集聚"的企业集群效应，最大限度地增加就业机会。它们提供了大量的人口，可以支持建设学校、医疗机构、公共交通和其他基础设施的投资。在这些场所，配套的生活设施都在步行和骑行距离之内，同时长距离出行可以选择公共交通，使健康有保障，强大的社区得以发展，从而助推当地企业的成功。

22 Making the best use of land means directing growth towards the most accessible and well-connected places, making the most efficient use of the existing and future public transport, walking and cycling networks. Integrating land use and transport in this way is essential not only to achieving the Mayor's target for 80 per cent of all journeys to be made by walking, cycling and public transport, but also to creating vibrant and active places and ensuring a compact and well-functioning city.

23 All options for using the city's land more effectively will need to be explored as London's growth continues, including the redevelopment of brownfield sites and the intensification of existing places, including in outer London. New and enhanced transport links will play an important role in allowing this to happen, unlocking homes and jobs growth in new areas and ensuring that new developments are not planned around car use.

24 As London develops, the Mayor's Good Growth by Design programme—which seeks to promote and deliver a better, more inclusive form of growth on behalf of all Londoners—will ensure that homes and other developments are of high quality. Existing green space designations will remain strong to protect the environment, and improvements to **green infrastructure** [14], biodiversity and other environmental factors, delivering more than 50 per cent green cover across London, will be important to help London become a **National Park City** [15].

25 London's distinctive character and heritage is why many people want to come to the city. London's heritage holds local and strategic significance for the city and for Londoners, and will be conserved and enhanced. As new developments are designed, the special features that Londoners value about a place, such as cultural, historic or natural elements, should be used positively to guide and stimulate growth, and create distinctive, attractive and cherished places.

26 Making the best use of land will allow the city to grow in a way that works for everyone. It will allow more high-quality homes and workspaces to be developed as London grows, while supporting local communities and creating new ones that can flourish in the future.

最大限度地利用土地，意味着将增长引向最易到达、交通最便利的地方，最有效地利用现有和未来的公共交通、步行和骑行网络。以这种方式整合土地利用和交通，不仅对实现市长提出的"80% 的行程由步行、骑行和公共交通完成"的目标至关重要，而且对创造充满活力和积极性的场所、确保形成紧凑和运作良好的城市也是必不可少的。

随着伦敦的不断发展，所有更有效利用城市土地的选项都需要被考虑，包括棕地的再开发和现有场所（包括伦敦外城）的集约化。新增的和改造的交通枢纽将发挥重要作用，带来新区域的住房释放和就业增长，并确保新的开发不会围绕汽车使用进行。

在伦敦的发展过程中，市长的"以设计实现美好增长"项目（Good Growth by Design programme）——旨在代表所有伦敦人推动和实现更好、更包容的增长形式——将确保住房和其他开发项目的高品质。现有的既定绿化空间将继续保护环境，而绿色基础设施、生物多样性和其他环境因素的改善，可以使整个伦敦的绿化覆盖率超过 50%，将帮助伦敦成为"国家公园城市"。

伦敦的独特气质和丰富遗存是许多人愿意来到这座城市的原因。对城市和市民来说，伦敦的建筑遗产具有地方性和战略性意义，应得到保护和加强。在设计新的开发项目时，应积极利用这些伦敦人珍视的地方特质，如文化、历史或自然元素，来引导和刺激发展，创造出独特的、迷人的、珍爱的场所。

充分利用土地将使这座城市以一种适合所有人的方式发展。随着伦敦的发展，它将允许开发更多高质量的住房和工作空间，同时支持当地社区，并创造新的社区，使其能够在未来蓬勃发展。

The Implementation Strategies of "GG2 Making the best use of land"

27 To create successful sustainable mixed-use places that make the best use of land, those involved in planning and development must:

A enable the development of brownfield land, particularly in Opportunity Areas, on surplus public sector land, and sites within and on the edge of town centres, as well as utilising small sites;

B prioritise sites which are well-connected by existing or planned public transport;

C proactively explore the potential to intensify the use of land to support additional homes and workspaces, promoting higher density development, particularly in locations that are well-connected to jobs, services, infrastructure and amenities by public transport, walking and cycling;

D applying a design-led approach to determine the optimum development capacity of sites;

E understand what is valued about existing places and use this as a catalyst for growth, renewal, and place-making, strengthening London's distinct and varied character;

F protect and enhance London's open spaces, including **the Green Belt** [16], Metropolitan Open Land, designated nature conservation sites and local spaces, and promote the creation of new green infrastructure and urban greening, including aiming to secure net biodiversity gains where possible;

G plan for good local walking, cycling and public transport connections to support a strategic target of 80 percent of all journeys using sustainable travel, enabling car-free lifestyles that allow an efficient use of land, as well as using new and enhanced public transport links to unlock growth;

H maximise opportunities to use infrastructure assets for more than one purpose, to make the best use of land and support efficient maintenance.

"GG2 充分利用土地"实施策略

为了创造可持续的混合用途场所，使土地得到充分利用，规划和开发的相关参与者必须：

A 能够开发棕地，特别是发展潜力地区（Opportunity Areas）、公共部门剩余土地（surplus public sector land）、城镇中心及边缘用地，同时利用小块用地；

B 优先考虑既有或规划中交通便利的用地；

C 积极探索土地集约利用的潜力，支持建造更多的住房和工作空间，推动更高密度的开发，特别是在那些通过公共交通、步行和骑行就能上班、使用服务、基础设施和生活设施的地区；

D 采用"以设计为主导"的方法，确定用地最适宜的开发容量；

E 了解既有场所的价值，并将其作为发展、更新和场所营造的催化剂，强化伦敦独特和多元的气质；

F 保护强化伦敦的开放空间，包括城市绿带、大都会区开阔地、指定的自然保护用地和地方性空间，并推动建立新的绿色基础设施和城市绿化，包括在可能的情况下确保生物多样性的净收益；

G 规划良好的步行、骑行和公共交通的地方网络，以支持实现"所有行程中 80% 使用可持续出行方式"的战略目标，实现无车生活方式，有效利用土地，利用新增与强化的公共交通线路来释放增长；

H 最大限度地创造机会将基础设施用于多个目的，充分利用土地并支持有效维护。

（文献来源：Mayor of London. The London Plan. https://www.london.gov.uk/what-we-do/planning/london-plan/new-london-plan/london-plan-2021. 杨慧、陈航、杨潇晗、黎赞翻译）

词汇 | Vocabulary

[1] Good Growth 美好增长。美好增长作为《伦敦规划 2021》未来的发展基调，强调了创造一个为所有伦敦人服务的城市。目标明确、政策可行、全民参与，均为了促进实现公平共享、人人可期的所有伦敦人的伦敦

[2] inclusive 包括的，包含的；规划专业中常意为"包容"

[3] underpin 巩固，支持

[4] decision-maker 决策者

[5] approach *v.* 靠近，邻近，近似，接近于；*n.* 方法，通道，路径

[6] inequality 不平等，不平均

[7] mixed-use development 混合用途开发项目

[8] Neighbourhood Plans 邻里计划

[9] thriving 欣欣向荣的，繁华的，兴旺的

[10] stakeholder 利益相关者

[11] sense of belonging 归属感

[12] buy-in（证券或商品交易中的）空头购入，买进（其他公司股份）；文中指对政策的同意，译为认同

[13] agglomeration 聚集

[14] green infrastructure 绿色基础设施，简称"GI"

[15] National Park City 国家公园城市

[16] the Green Belt 城市绿带

练习与思考 | Comprehension Exercise

1. A city that is planned well can improve as it grows.

理解并翻译该句，使其符合城市规划专业术语并符合政府公文语境。

2. Taking advantage of the knowledge and experience of local people will help to shape London's growth, creating a thriving city that works better for all Londoners.

详细分析句子结构，理解句中动词的多种形式。

课后延伸 | Reading Material

美好增长 · 伦敦的未来（续）| Good Growth · Planning London's Future（part 2）

GG3 Creating a healthy city

1 The mental and physical health of Londoners is, to a large extent, determined by the environment in which they live. Transport, housing, education, income, working conditions, unemployment, air quality, green space, climate change and social and community networks can have a greater influence on health than healthcare provision or genetics. Many of these determinants of health can be shaped by the planning system, and local authorities are accordingly responsible for planning and public health.

2 As set out in the Mayor's *Health Inequalities Strategy*, the scale of London's health inequalities is great and the need to reduce them is urgent. Healthy life expectancy is lower in more deprived areas, and the differences between parts of London is stark – more than 15 years for men and almost 19 years for women. London's ongoing growth provides an opportunity to reduce these inequalities. Delivering Good Growth will involve prioritising health in all of London's planning decisions, including through design that supports health outcomes, and the assessment and mitigation of any potential adverse impacts of development proposals on health and health inequality.

3 The causes of London's health problems are wide-ranging. Many of London's major health problems are related to inactivity. Currently only 34 percent of Londoners report doing the 20 minutes of active travel each day that can help them to stay healthy, but good planning can help them to build this into their daily routine. Access to green and open spaces, including waterways, can improve health, but access and quality varies widely across the city. Excessive housing costs or living in a home that is damp, too hot or too cold can have serious health impacts. A healthy food environment and access to healthy food is vital for good health. Good planning can help address all these issues.

4 The *Healthy Streets Approach* outlined in this plan puts improving health and reducing health inequalities at the heart of planning London's public space. It will tackle London's inactivity crisis, improve air quality and reduce the other health impacts of living in a car-dominated city by planning street networks that work well for people on foot and on bikes, and providing public transport networks that are attractive alternatives to car use. It will also ensure that streets become more social spaces.

GG3 创建健康城市

伦敦人的身心健康在很大程度上是由生活环境决定的。交通、住房、教育、收入、工作条件、失业率、空气质量、绿化空间、气候变化以及社会和社区网络对健康的影响可能比医疗服务或遗传因素更大。这些决定健康的因素中，有许多可以由规划系统来决定，因此地方当局对规划和公共卫生负有责任。

正如市长在《健康不平等现象的应对策略》（*Health Inequalities Strategy*）中所述，伦敦的健康不平等现象非常严重，迫切需要改善这种现象。贫困地区的预期健康寿命较低，而伦敦各地区之间的差异也非常明显——男性超过 15 年而女性接近 19 年。伦敦的持续发展为改善这些不平等现象提供了机遇。实现"美好增长"将在伦敦所有规划决策中优先考虑健康因素，支持健康成果设计，对开发方案进行评估，排除 / 缓和与健康不平等有关的任何潜在不利影响。

造成伦敦健康问题的原因是多方面的。伦敦的许多主要健康问题都与低运动量有关。目前，只有 34% 的伦敦人表示每天进行 20 分钟有助于保持健康的积极出行，但良好的规划可以帮助人们改变日常习惯。可到达的绿化和开放空间，包括水道，可以改善健康状况，但在整个城市中，可达性及环境品质差异很大。过高的住房成本或居住在潮湿、过热或过冷的住房中会对健康产生严重影响。健康的食物环境和获得健康食物的机会也对保持健康至关重要。良好的规划可以帮助解决所有这些问题。

本文件中概述的《健康街道方法》（*Healthy Streets Approach*）将改善健康状况和减少健康不平等现象作为伦敦公共空间规划的核心，将处理伦敦的低运动量危机，改善空气质量，减少生活中因汽车主导而带来的其他健康问题。采用的方法是规划适合步行和骑行的街道网络，提供有吸引力的公共交通网络来替代汽车，确保街道成为更好的社交空间。

5   The social and environmental causes of ill-health are numerous and complex, and the people who are most affected by London's health inequalities tend also to be affected by other forms of inequality. Creating a healthy city with reduced health inequalities will make London fairer for everyone. The Mayor plays a pivotal role in bringing together a diverse range of stakeholders from service providers, boroughs, communities and the private sector in order to provide a more integrated approach to promoting a healthy city and reducing health inequalities. The Mayor will co-ordinate investment and focus regeneration initiatives in those parts of London most affected by inequalities, including health inequalities.

The Implementation Strategies of "GG3 Creating a healthy city"

6   To improve Londoners' health and reduce health inequalities, those involved in planning and development must:

A ensure that the wider determinants of health are addressed in an integrated and coordinated way, taking a systematic approach to improving the mental and physical health of all Londoners and reducing health inequalities;

B promote more active and healthy lives for all Londoners and enable them to make healthy choices;

C use the Healthy Streets Approach to prioritise health in all planning decisions;

D assess the potential impacts of development proposals and Development Plans on the mental and physical health and wellbeing of communities, in order to mitigate any potential negative impacts, maximise potential positive impacts, and help reduce health inequalities, for example through the use of Health Impact Assessments;

E plan for appropriate health and care infrastructure to address the needs of London's changing and growing population;

F seek to improve London's air quality, reduce public exposure to poor air quality and minimise inequalities in levels of exposure to air pollution;

G plan for improved access to and quality of green spaces, the provision of new green infrastructure, and spaces for play, recreation and sports;

H ensure that new buildings are well-insulated and sufficiently ventilated to avoid the health problems associated with damp, heat and cold.

导致不良健康状况的社会和环境原因很多，也很复杂，受伦敦健康不平等现象影响最大的人群往往也更容易受到其他不平等的影响。创造一个健康城市，减少健康不平等，将使伦敦对每个人都更加公平。市长将在促进建立健康城市和减少健康不平等方面发挥关键性作用，包括聚集服务供应商、行政管理、社区和私营部门等不同的利益相关者，提供更综合的实现方法。市长将协调投资，使重点更新项目着力于那些受不平等影响最严重的区域（包括健康不平等）。

"GG3 创建健康城市"实施策略

为了改善伦敦人的健康状况并减少健康不平等现象，规划和开发的相关参与者必须：

A 确保以综合与协调的方式确定广泛的健康因素，采取系统性的方法改善所有伦敦人的身心健康并减少健康不平等现象；

B 向所有伦敦人推广更积极更健康的生活方式，使他们能够作出有利健康的选择；

C 采用《健康街道方法》，在所有规划决策中优先考虑健康问题；

D 评估开发方案和开发建设规划对社区健康与福祉的潜在影响，减轻任何潜在的负面影响，最大限度地发挥积极影响，并帮助减少健康不平等现象，如开发项目必须通过健康影响评估；

E 规划适当的医疗保健基础设施，以满足伦敦不断变化和增长的人口的需求；

F 改善伦敦的空气质量，减少公众暴露在不良空气中的机会，尽量降低暴露于污染空气中的不平等水平；

G 规划改善绿化空间的可达性和品质，提供新的绿色基础设施，以及游戏、娱乐和运动空间；

H 确保新建筑有良好的隔热和充分的通风，避免潮湿、炎热和寒冷产生的健康问题。

I seek to create a healthy food environment, increasing the availability of healthy food and restricting unhealthy options.

GG4 Delivering the homes Londoners need

7　Few things are as important as a suitable home, but for many Londoners the type of home they want, and should reasonably be able to expect, is out of reach. In 2016, the gap between average house prices in London and the rest of the country reached a record high, and the private rental cost of a one-bedroom home in London is now more than the average for a three-bedroom home in any other English region. A housing market that only works for the very wealthy does not work for London.

8　The state of London's housing market has implications for the makeup and diversity of the city. Affordable housing is central to allowing Londoners of all means and backgrounds to play their part in community life. Providing a range of high quality, well-designed, accessible homes is important to delivering Good Growth, ensuring that London remains a mixed and inclusive place in which people have a choice about where to live. The failure to provide sufficient numbers of new homes to meet London's need for affordable, market and specialist housing has given rise to a range of negative social, economic and environmental consequences, including: worsening housing affordability issues, overcrowding, reduced labour market mobility, staff retention issues and longer commuting patterns.

9　The lack of supply of the homes that Londoners need has played a significant role in London's housing crisis. The *2017 London Strategic Housing Market Assessment* has identified a significant overall need for housing, and for affordable housing in particular. London needs 66,000 new homes each year, for at least twenty years, and evidence suggests that 43,000 of them should be genuinely affordable if the needs of Londoners are to be met. This supports the Mayor's strategic target of 50 percent of all new homes being genuinely affordable, which is based on viability evidence.

10　The *London Plan 2021* is able to look across the city to plan for the housing needs of all Londoners, treating London as a single housing market in a way that is not possible at a local level. In partnership with boroughs, the Mayor has undertaken a "Strategic Housing Land Availability Assessment" to identify

笔者努力创造健康的食物环境，增加健康食物的供应，限制不健康的选项。

GG4 提供伦敦人需要的住房

住房问题是重中之重。对于许多伦敦人来说，他们想要的住宅类型是遥不可及的。2016 年，伦敦平均住宅价格与全国其他地区的差距创下历史新高，伦敦一居室住宅的私人租赁成本已经超过了英国其他地区三居室住宅的均价。只适合极富人群的住房市场对伦敦来说并不合理。

伦敦住房市场的状况对这座城市的性格和多元化有重要影响。经济适用住宅是让所有不同经济能力和背景的伦敦人可以在社区生活中发挥作用的核心。提供一系列高质量、设计精良、交通便利的住宅对实现"美好增长"至关重要，伦敦仍然是一个人们可以选择在哪里生活的多元而包容的地方。不能提供充足数量的新建住宅来满足伦敦对经济适用住宅、市场住房及专营住房的需求，已经导致了一系列负面的社会、经济和环境后果，如住宅经济适用性问题的恶化、过度拥挤、劳动力市场流动性降低、员工留用问题和更耗时的通勤模式。

住宅供应不足是伦敦住宅危机的一个重要原因。《2017 年伦敦战略性住房市场评估》已确认了对住宅，特别是经济适用住宅的巨大需求。伦敦在未来 20 年中每年需要 6.6 万套新建住房，其中 4.3 万套应该是真正经济适用的住宅。这支持了市长的战略目标，即所有新房中 50% 是真正经济适用的。

《伦敦规划 2021》着眼于整个城市，为所有伦敦人的住房需求进行规划，将伦敦作为一个完整的住房市场，这在地市级本是不可能实现的。市长与各自治市镇合作，进行了"战略性住宅用地可用性评估"，以确定伦敦所需住宅的交付地点。

where the homes London needs can be delivered. Ten-year housing targets have been established for every borough, alongside Opportunity Area plans for longer-term delivery where the potential for new homes is especially high. Boroughs can rely on these targets when developing their Development Plan Documents and are not required to take account of nationally-derived local-level need figures.

11 To meet the growing need, London must seek to deliver new homes through a wide range of development options. Reusing large brownfield sites will remain crucial, although vacant plots are now scarce, and the scale and complexity of large former industrial sites makes delivery slow. Small sites in a range of locations can be developed more quickly, and enable smaller builders to enter the market. Building more housing as part of the development of town centres will also be important, providing homes in well-connected places that will help to sustain local communities.

12 The homebuilding industry itself also needs greater diversity to reduce our reliance on a small number of large private developers. New and innovative approaches to development, including Build to Rent, community-led housing, and self- and custom-build, will all need to play a role, and more of our new homes will need to be built using precision-manufacturing. Alongside this, there will need to be a greater emphasis on the role councils and housing associations play in building genuinely affordable homes.

13 There are a range of other measures that have an impact on the availability of homes. For example, existing homes must not be left empty, and have the potential to be brought back into use as affordable housing, and boroughs should use all the tools at their disposal to ensure that homes are actually built after planning permissions are granted.

14 Delivering the housing London needs will be a huge challenge that will require everyone involved in the housing market to work together. Along with the London Housing Strategy, this *London Plan 2021* establishes the framework that will make this possible, helping to make London a city that everyone who wants to can call home.

每个市镇都制定了十年住宅目标，同时制订了"发展潜力地区计划"，以便在新房潜力特别大的地方长期提供住房。各自治市镇在制订开发计划时可以依据这些目标，而不需要考虑基于全国性调查而得到的有关本地需求的数据。

为了满足日益增长的需求，伦敦必须通过广泛的开发方案来提供新的住宅，重新利用大型棕地仍将是至关重要的。规模较大的前工业用地往往情况复杂，使得交付速度缓慢，小型地块则便于更快地开发，使小型建筑商能够进入市场。住房是城镇中心发展的重要部分，在交通便利的地方提供住房将有助于维持当地社区的发展。

住房建筑业本身也需要更加多样化，减少对少数大型私营开发商的依赖。创新的发展方式，如建房出租（Build to Rent）、社区住宅（community-led housing）、自建与定制住宅（self- and custom-build），都需要发挥作用，同时需要采用精确的建造方式。除此之外，还需要更加强调议会和住房协会的作用。

其他措施也将对住宅供应产生影响。例如，既有住宅不应空置，尽可能重新被用作经济适用住房，利用一切可利用的工具来确保住房在获得规划许可后被真正建造。

提供伦敦所需的住宅将是一个巨大的挑战，需要所有住房市场的参与者共同努力。《伦敦规划 2021》将与"伦敦住房战略"一起，建立一个可行的框架，帮助伦敦成为一座每个人都想称之为家的城市。

The Implementation of Strategies of "GG4 Delivering the homes Londoners need"

15. To create a housing market that works better for all Londoners, those involved in planning and development must:

A ensure that more homes are delivered;

B support the delivery of the strategic target of 50 per cent of all new homes being genuinely affordable;

C create mixed and inclusive communities, with good quality homes that meet high standards of design and provide for identified needs, including for specialist housing;

D identify and allocate a range of sites to deliver housing locally, supporting skilled precision-manufacturing that can increase the rate of building, and planning for all necessary supporting infrastructure from the outset;

E establish ambitious and achievable build-out rates at the planning stage, incentivising build-out milestones to help ensure that homes are built quickly and to reduce the likelihood of permissions being sought to sell land at a higher value.

"GG4 提供伦敦人需要的住房"实施策略

为了创造更适合所有伦敦人的住房市场，规划和开发的相关参与者必须：

A 确保提供更多住房；

B 支持实现所有新建住宅中 50% 为真正经济适用住宅的战略目标；

C 建立多元、包容的社区，提供符合高标准设计的优质住宅，并满足包括专营住房在内的特定需求；

D 确定并分配一系列用地以支持住宅建设，支持能够提高建房率的精确建造方式，并在规划初期确定必要的配套基础设施；

E 在规划阶段制定雄心勃勃又可实现的扩建率，激励扩建里程碑，以帮助快速建造住房并降低高价出售土地的可能性。

（文献来源：Mayor of London. The London Plan. https://www.london.gov.uk/what-we-do/planning/london-plan/new-london-plan/london-plan-2021. 杨慧、陈航、杨潇晗、黎赟翻译）

第 3 课　可持续发展

Lesson 3　Sustainable Development

导读 | Introduction

全球城市五大趋势 | Five Trends in Global Cities

在 2050 年以及更远的未来，我们的城市会变成什么样子？城市会不会过于拥挤？我们的生活会不会过于匆忙？我们居住的社区在极端天气下还适宜居住么？

城市正在快速发展，给我们带来了新的挑战和机遇，同时也需要新的工具与方法。大规模的城市化意味着我们在创造和实现城市愿景方面的共同成功将影响我们每个人，甚至是那些生活在城市之外的人。

全球城市五大趋势如表 3-1 所示。

表 3-1　全球城市五大趋势

	1. 全球城市化进程	2016 年 54% 的人居住于城镇； 2050 年 66% 的人居住于城镇
	2. 气候紧急情况	对城市的危害包括降雨、洪涝、干旱和高温等

续表

	3. 世界人口增长登顶	预计到 2050 年，55 个国家和地区，特别是欧洲国家的人口将会下降。与此同时，人口也在老龄化。经济增长时期的投资将影响未来几十年的生活质量
	4. 指数级增长的技术	将创造超出想象的机会。面对面的互动更加自由，更加强调身体的体验
	5. 新型经济	活跃的城市环境和机遇塑造了共享经济、循环经济、体验经济与创新经济

为了适应增长、建设可持续的城市，许多政府对既有城区采取了致密化（densification）[1] 的规划战略，尽管往往伴随着有针对性的城市扩张。致密化经常发生在就业岗位集中和服务业发达的地方，也会发生在以前的工业区，以及带有大量闲置产能的既有或新建的公共交通设施附近。

城市致密化目标：

· 优化城市基础设施与服务；

· 尽量减少资源和能源消耗；

· 促进健康旅行；

· 提高社会凝聚力；

· 减少气候影响。

预备知识 | Preliminary Knowledge

碳达峰碳中和目标 | The Carbon Peak and Carbon Neutrality Target

2020 年 9 月 22 日，中国国家主席习近平在第七十五届联合国大会一般性辩论上宣布："中国将提高国家自主贡献力度，采取更加有力的政策和措施，二氧化碳排放力争于 2030 年前达到峰值，努力争取 2060 年前实现碳中和。"中国碳达峰碳中和目标（即"双碳"目标）的提出在国内和国际社会引发关注，也成为我国"十四五"污染防治攻坚战的主攻目标。

碳达峰碳中和目标（the Carbon Peak and Carbon Neutrality Target）：碳达峰即我国承诺在 2030 年前，二氧化碳排放量不再增长，达到峰值后逐步减少；碳中和是指到 2060 年，针对排放的二氧化碳，要采取植树、节能减排等各种方式全部抵消。通俗来讲，碳达峰指二氧化碳排放量在某一年达到了最大值，之后进入下降阶段；碳中和则指一段时间内，特定组织或整个社会活动产生的二氧化碳，通过植树造林、海洋吸收、工程封存等自然、人为手段被吸收和抵消掉，实现人类活动二氧化碳相对"零排放"。

目前，国际社会对于脱碳十分关注。作为世界上人口最多的城市，东京在控制碳排放方面面临着挑战。麦肯锡咨询公司从技术和投资的角度，为日本主要行业分析了目前最可行、最经济的脱碳之路，分析了日本 2030 年前和 2030 年到 2050 年的可行的脱碳之路，并给出不同时期具体行动和资金筹措的建议。

《日本 2050 年碳中和之路》就是麦肯锡进行的以 ESG 为导向的咨询分析。

ESG 是英文 Environmental（环境标准）、Social（社会标准）和 Governance（治理）的缩写，是一种关注企业环境、社会、治理绩效而非财务绩效的投资理念和企业评价标准。基于 ESG 评价，投资者可以通过观测企业 ESG 绩效、评估其投资行为和企业（投资对象）在促进经济可持续发展、履行社会责任等方面的贡献。ESG 评价是开展业务中重要的部分，并与创造价值息息相关。

课文讲解 | Text Explanation

How Japan Could Reach Carbon Neutrality by 2050

Developing Japan's cost-optimal pathway

1　There are many paths Japan could take to reach the **net-zero target**[2]. We modeled a pathway that from a central planner's perspective would be considered "**societally cost-optimal**[3]," with a **social discount rate**[4] of 4 percent for all investments. We optimized cost for the whole system, including every sector, and for the entire time horizon, from 2017 to 2050.

2　We ran a **bottom-up analysis**[5] using two principal proprietary models.

3　McKinsey Decarbonization Pathway Optimizer (DPO): A model with more than 600 technologies in 75 segments. Each technology is attached to a business case, including investment and operating-cost components, emissions impact, and energy consumption.

4　McKinsey Power Model (MPM): A power-system model that simulates electricity supply and demand on an hourly basis to arrive at a cost-optimized power-generation technology mix.

5　Our pathway is not a prediction of what will happen under current policy, social and technological conditions. Nor does it account for all the unique situations faced by each company and individual in Japan. Our intent is to help inform the planning efforts of policy makers and business leaders, demonstrate the technical **feasibility**[6] of achieving Japan's emissions-reduction targets, and explore the implications of the changes that would be required.

The cost-optimal pathway through 2030

6　For Japan to reach its 46 percent emissions-reduction target by 2030, it would need to eliminate about 500

日本 2050 年碳中和之路

发展成本最优的脱碳之路

日本有很多途径可以实现零排放的目标，我们从中央政府的角度模拟了一个"社会成本最优"的减排路径。在该路径下，所有投资的社会贴现率仅为 4%。我们优化了每个部门以及从 2017 年到 2050 年的时间范围内的整个系统的成本。

我们使用了两个自有模型进行了自下而上的分析。

麦肯锡最优脱碳路径分析模型（DPO）：在 75 个细分市场中拥有 600 多项技术的模型。每项技术都与一个商业案例相关联，包括投资和运营成本、温室气体排放影响和能源消耗。

麦肯锡电力模型（MPM）：该模型可以以每小时为基础模拟电力供应和需求，并最终得出成本最优化的集成发电技术。

我们提出的脱碳路径并不是基于目前政策、社会环境和技术条件对未来的预测，也没有特殊针对某一个日本公司或者个人所面临的情况。我们的目的是为政策制定者和商界领袖提供减排脱碳规划工作的信息，证明实现日本减排目标的技术可行性，并探索在这一过程中政策、技术变革所带来的影响。

到 2030 年的成本最优路径

日本想要在 2030 年时达到减排 46% 的目标意味着其需要清除 500 吨二氧化碳当量（tCO$_{2e}$）。

tCO_{2e}. This could be done at an average cost savings of \$34 per metric ton of carbon dioxide equivalent (tCO_{2e}) over the next decade because the required technologies are already mature.

7   The **buildings sector**[7] would be the fastest to **decarbonize**[8], reducing emissions 55 percent, at a cost savings of \$57 per tCO_{2e}, by installing better insulation and switching to electric heat pumps instead of fossil-fuel boilers.

8   The power sector would be the second fastest to decarbonize, reducing emissions 42 percent by 2030. This could be achieved by replacing coal power plants with combined-cycle gas turbines (CCGT), restarting nuclear plants closed after the 2011 Fukushima accident, and expanding offshore wind and solar-power capacity—measures that would generate an average cost savings of \$18 per tCO_{2e} through 2030.

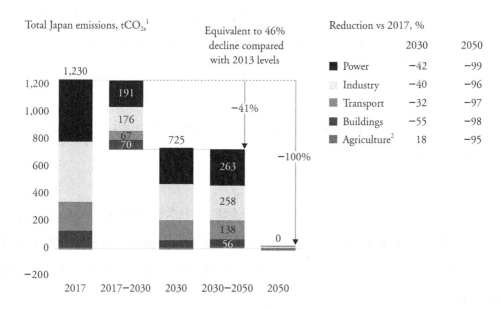

In Japan's decarbonization pathway, buildings and power have started first.

1 Metric megatons of carbon dioxide equivalent.

2 Including land use, land-use change, and forestry(LULUCF) or natural carbon sequestration from land use and forests.

(Source: Japan Ministry of Economy "Trade I and Industry; McKinsey Analysis")

这在未来十年里可以实现。由于所需的技术已经成熟，每减排 1 吨二氧化碳当量（tCO_{2e}）平均可以节省 34 美元的成本。

通过安装更好的绝缘材料并改用电热泵取代化石燃料锅炉，建筑部门将会是最快实现脱碳的部门。到 2030 年，建筑部门预计减少 55% 的排放量，且每减排 1 吨二氧化碳当量可节省 57 美元的成本。

电力部门将是脱碳速度位列第二的部门，到 2030 年预计将减少 42% 的排放量。通过用联合循环燃气轮机（CCGT）取代燃煤电厂、重新启动 2011 年福岛核事故后关闭的核电厂以及扩张海上风能和太阳能发电能力，日本电力部门到 2030 年可实现每减排 1 吨二氧化碳当量平均节省 18 美元的成本。

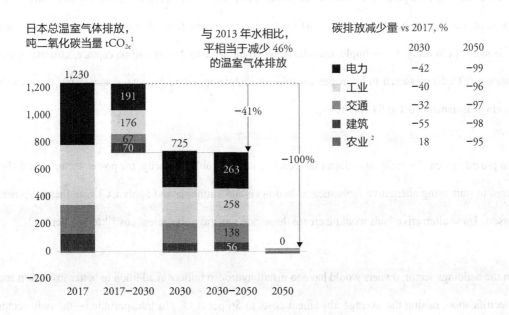

在日本的脱碳之路上，建筑和电力部门已首先开始

---

1 吨二氧化碳当量。

2 包含土地利用、土地利用改变和林业（LULUCF）或者来源于土地利用和森林的自然碳封存。

（来源：日本经济产业省《麦肯锡分析报告》）

9　Industry remains a difficult sector to decarbonize because of its high-temperature processes and carbon-intensive feedstocks. But through demand reduction, switching from oil to gas in industrial boilers, and using heat pumps for low-temperature processes, industry emissions could be reduced 40 percent by 2030 at an average cost savings of $20 per tCO_{2e}.

10　The transportation sector would see the smallest drop in emissions by 2030, at just 32 percent. Although the technology to decarbonize transportation is mature, it will take a while to **ramp up** [9] battery electric vehicle (BEV) manufacturing, scale the charging infrastructure, and motivate passenger-car and light-commercial-truck owners to make the switch. Through 2030, the average abatement cost savings would be $49 per tCO_{2e}.

The cost-optimal pathway from 2030 to 2050

11　Now comes the hard part. After most of the easier, more affordable decarbonization solutions have been implemented, Japan would have to **resort to** [10] more expensive technologies to reduce its remaining emissions. In industry, for example, manufacturers would have to deploy carbon capture, utilization, and storage (CCUS) or switch to hydrogen for mid- to high-temperature processes, increasing the sector's average abatement cost to $41 per tCO_{2e}.

12　To provide green electricity after Japan has reached its renewables capacity, the power sector would also need to start using alternative fuels such as hydrogen and ammonia and apply CCUS to thermal-power assets. These alternative fuels would increase the sector's average **abatement cost** [11] to $81 per tCO_{2e}.

13　In the buildings sector, owners would have to install hydrogen boilers in addition to better insulation and electrification, raising the average abatement costs to $6 per tCO_{2e}. In transportation—the only sector that would see overall cost savings—Japan would have to supplement the electrification of passenger and light-duty vehicles by switching to fuel cell technology for long-haul trucks and hydrogen and biofuels for aircraft and ships. These changes would result in an average abatement cost savings of $62 per tCO_{2e}.

由于涉及高温工艺和碳密集型原料，工业部门仍然是一个难以脱碳的行业。但是，通过减少需求、将工业锅炉使用的石油转变为天然气以及在低温过程中使用热泵，日本到 2030 年可以减少 40% 的工业排放量，平均每吨二氧化碳当量的减排成本节省 20 美元。

到 2030 年，日本交通部门的排放量下降幅度将会是五部门中最少的，仅为 32%。尽管脱碳技术在交通领域已经成熟，但要加大纯电动汽车（BEV）的制造力度、扩大电动汽车充电基础设施的规模并激励乘用车和轻型商用卡车车主从燃油汽车转向电动汽车仍需要一段时间。到 2030 年，日本交通部门平均每减排一吨二氧化碳当量可节省 49 美元的成本。

2030 年到 2050 年的成本最优路径

2030 年后日本将面临更具挑战性的脱碳之路。在大多数更容易、成本更低的脱碳解决方案实施后，日本将不得不求助于更昂贵的技术来减少其剩余的排放量。例如，在工业中，制造商将不得不利用碳捕获、利用和储存（CCUS 策略）或在中高温过程中改用氢气，从而使该行业的平均减排成本增加到每吨二氧化碳当量 41 美元。

在日本实现可再生能源产能后，电力部门为了能提供绿色电力，还需要开始使用氢和氨等替代燃料，并将 CCUS 策略应用于火电资产。这些替代燃料将使该行业的平均减排成本增加到每吨二氧化碳当量 81 美元。

2030 年后，在日本建筑部门，除了更好地使用绝缘材料和实现电气化外，业主还必须安装氢气锅炉，这一措施将会把平均减排成本提高到每吨二氧化碳当量 6 美元。唯一能实现总体成本节约的部门是交通运输。除了乘用车和轻型车辆的电气化外，日本还必须通过在长途卡车上使用燃料电池技术以及在飞机和船舶运输上采用氢燃料和生物燃料以达到最终零排放目标。这些技术变化将使得平均每吨二氧化碳当量的减排成本节省 62 美元。

14 As a result of these efforts, the average abatement cost for all sectors together would rise to $36 per tCO_{2e} by 2050, a significant increase from the cost savings of $34 per tCO_{2e} through 2030. During the **transition** [12] , **primary energy** [13] inputs and energy consumption would drop as activity reduced in line with Japan's declining population and increased process and fuel efficiencies. Oil and coal consumption would disappear, while **renewables** [14] and **clean hydrogen and ammonia** [15] would become the primary energy supplies. However, natural gas would remain a part of the mix to supplement demand that could not entirely be fulfilled by renewables.

Highlights of required actions

15 The magnitude of achieving net-zero is illustrated by some of the actions that would need to be taken.

16 Power: Solar and wind capacity would need to increase **threefold** [16] , to 275 gigawatts (GW) by 2050. Unabated coal-fired power generation would be shut down by 2030.

17 Industry: Because electrification can't generate the heat for mid- to high-temperature manufacturing processes, this sector would need to rely on hydrogen to reduce 21 percent of its emissions and CCUS for an 18 percent reduction. Japan **would have to** [17] establish a hydrogen supply chain and invest in expensive CCUS technology.

18 Transportation: To meet the 2030 emissions-reduction target, 90 percent of new cars, trucks, and buses sold in Japan by 2030 would need to be BEVs. The auto industry would have to ramp up BEV production, and cities would need to install the necessary **infrastructure** [18] .

19 Buildings: Because better insulation and electrification of space heating and cooking can't eliminate all building emissions, 10 percent of the total energy use would have to be hydrogen, **reinforcing** [19] Japan's need to establish a robust hydrogen supply chain.

由于这些努力，到 2050 年，日本所有部门的平均减排成本将升至每吨二氧化碳当量 36 美元，与 2030 年的平均减排成本每吨二氧化碳当量 34 美元相比大幅增加。转型期间，一次能源投入和能源消耗将随着日本人口的减少、日常活动的减少以及能源处理流程和燃料效率的提高而减少。石油和煤炭消费将消失，而可再生能源和清洁的氢氨能源将成为主要能源供应来源。然而，天然气仍将作为补充能源的一部分以满足可再生能源无法完全满足的需求。

应采取的行动要点

实现净零的重要性就体现在这些应采取的行动中。

能源：到 2050 年，太阳能和风能的发电能力应增加两倍，达到 275 千兆瓦。到 2030 年，将停止燃煤发电。

工业：由于电气化不能为中高温制造工艺提供热量，因此该领域需要依靠氢气能源来减少 21% 的排放，依靠 CCUS 策略减少 18% 的排放。日本将不得不建立一条氢气供应链，并投资昂贵的 CCUS 技术。

交通：为了实现 2030 年的减排目标，到 2030 年，日本国内销售的 90% 的新汽车、卡车和公共汽车都必须是纯电动汽车。汽车行业将不得不提高纯电动汽车的产量，城市也将需要安装必要的基础设施。

建筑：即使改善了建筑隔热材料并完全采用电气化加热和烹饪方式，也不能完全消除建筑物的碳排放，因此，总的使用能源中的 10% 必须更换为氢气。也因此日本需要建立健全的氢气供应链。

Flow of primary energy supply to final energy consumption

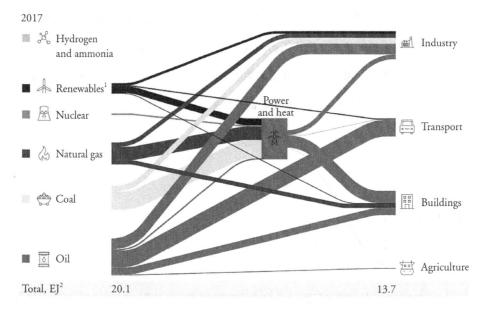

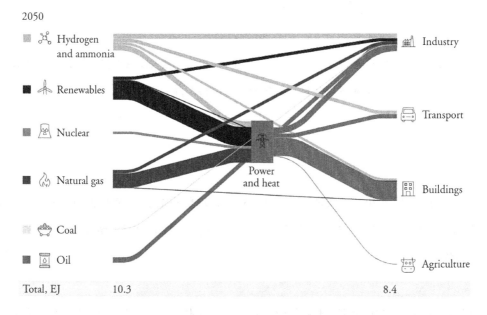

Renewables and low-carbon hydrogen and ammonia will become the major primary energy supply, but natural gas will remain significant.

Note: Depending on domestic production vs import source share of hydrogen and ammonia, hydrogen and ammonia in primary energy supply(import) can decrease and be replaced with natural gas and second energy transformation into hydrogen and ammonia.

1 Includes hydroelectric power, biomass, solar, wind, geothermal.

2 Exajoules.

(Source: Japan Ministry of Economy "Trade and Industry; McKinsey Analysis")

一次能源到终端能源消耗的能源流

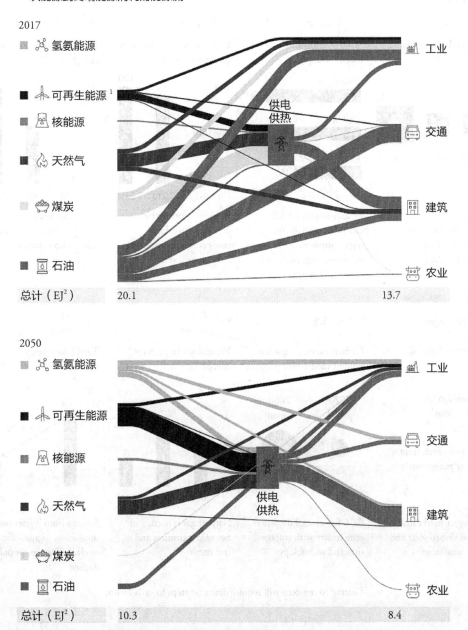

可再生能源和低碳的氢氨能源将成为主要的能源供应来源，但天然气仍然是重要的能源之一。

注：根据氢氨能源国内生产和进口来源的比例，一次能源中的氢氨能源可能会减少并被天然气替代，进而在二次能源转换过程中变为氢氨能源。
1 包含水力发电、生物质、太阳能、风和地热。
2 艾焦。
（来源：日本经济产业省《麦肯锡分析报告》）

Sectoral highlights

Power

Solar and wind installed capacity, GW[1]

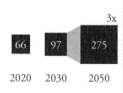

| 66 | 97 | 275 |

3x

2020 2030 2050

3x uptake in 2030-2050 as cost declines and land constraints debottlenecked

Industry

Share of emissions reduction lever, %

Hydrogen ████████████ 21

CCUS[2] ██████████ 18

Electrification ████████ 15

Hydrogen and CCUS crucial , especially for high-temperature heat subsectors (eg, steel, cement, chemical)

Transport

BEV[3] share in annual new passenger car sales, %

100 ─
90 ──── 100
180x
0 ── 0.5
2020 2030 2050

Massive ramp-up by 2030 due to relative cost competitiveness vs alternative levers

Buildings

Hydrogen share in final energy use, %

10 ─── 10
6
0 ── 0
2020 2030 2050

Hydrogen needed due to supply constraint of cheap zero-carbon electricity

Thematic highlights

Hydrogen

Hydrogen demand, 2050

22 million metric tons[4] per year

31% of which from power generation

Hydrogen-fired capacity needed due to solar and wind constraint

CCUS

Carbon capture capacity, million metric tons per year

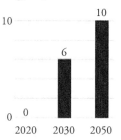

2020 global   2050 Japan

40   173

>4x

CCUS is crucial despite immaturity with storage sites and technology

Primary energy

Natural gas in primary energy, EJ[5]

10 ─
7
5
3
0 ──
2020 2030 2050

Natural gas is needed in net-zero transition and end result

Final energy

Total final energy use , EJ

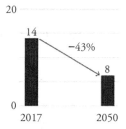

20 ─
14
−43%
8
0 ──
2017   2050

Savings from higher building insulation, higher efficiency of electrification, population decline

Getting to net-zero will require dramatic steps in each sector.

---

1 GW(Gigawatts): 1GW=109W

2 CCUS: Carbon capture use and storage.

3 BEV: Battery electric vehicle

4 1 metric ton = 2,205 pounds.

5 EJ(Exajoules): 1 EJ=10^{18}J

(Source: Japan Ministry of Economy "Trade and Industry; McKinsey Analysis")

行业重点

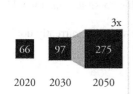

能源

太阳能和风能的发电能力（GW[1]）

3x

66　97　275

2020　2030　2050

随着成本的下降和土地限制的解除，2030—2050年的增长率将提升3倍

工业

减排量杠杆 (%)

氢　21

CCUS[2]　18

电气化　15

氢和CCUS至关重要，特别是对高温行业（如钢铁、水泥、化工）

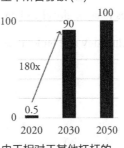

交通

BEV[3] 在年度新乘用车销量中所占份额 (%)

100　90　100

180x

0.5

2020　2030　2050

由于相对于其他杠杆的成本竞争力，到2030年BEV的份额将大幅提高

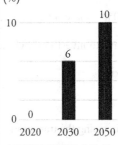

建筑

氢在能源使用中所占份额 (%)

10　10

6

0

2020　2030　2050

由于廉价零碳电力的供应限制，将更需要氢能

主题重点

氢

氢能需求
2050

每年2200
万吨[4]

其中31%用于发电

由于太阳能和风能的限制，需要用氢燃料来发电

CCUS

碳捕获能力
百万吨 / 每年

2020　2050
全球　日本

40　173

>4x

CCUS是至关重要的，尽管碳存储地点和技术尚不成熟

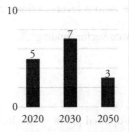

一次能源

一次能源中的天然气消耗 (EJ[5])

10

5　7　3

2020　2030　2050

在净零过渡期和最终阶段中都需要天然气参与

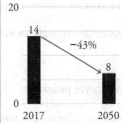

终端能源

终端能源使用总计 (EJ)

20

14

−43%

8

2017　2050

更高的建筑材料隔热性，更高的电气化效率，人口下降都节省了能源使用量

要实现净零排放，每个部门都需要采取重大措施

---

1 吉瓦，1 吉瓦等于 10 亿瓦特
2 碳捕获、碳利用和碳储存
3 纯电动汽车
4 吨，1 吨 =2205 磅
5 艾焦（即 10^{18} 焦）

（数据来源：日本经济产业省《麦肯锡分析报告》）

20　To make these changes, Japan would need 22 million tons of hydrogen a year by 2050, up significantly from the less than two million tons required today. The power sector would consume 31 percent of that hydrogen for power generation. CCUS technology would need to capture 173 million tons of CO_2 by 2050—more than four times today's global carbon-capture volume. To make this possible, Japan would have to invest in CCUS technology and build a vast network of carbon storage sites that currently don't exist.

21　Financing the transition

On our cost-optimal pathway, reaching net-zero would require a total investment of $10 trillion by 2050, or $330 billion annually. Of that investment, $8 trillion would come from **redirecting funds** [20] that would have been invested in **incumbent technologies** [21]. An additional $2 trillion—an average of $70 billion annually, or 1 to 2 percent of the country's GDP—would be needed to cover the higher net cost of the decarbonizing technologies and infrastructures that are more expensive to implement, such as an expanded power grid, BEV charging stations, and pipelines for hydrogen and ammonia transmission. The government can utilize various levers to **finance** [22] the transition, such as **subsidies** [23], carbon taxes, special investments, or **private-sector funding** [24].

22　Alternative pathways

While developing our cost-optimal pathway, we looked at a few other ways that Japan could reach net-zero because so many variables are in play.

23　In the first alternative pathway, Japan would push for renewables to constitute almost 80 percent of its power generation rather than 60 percent. Achieving this would require addressing geological and social constraints, such as confronting **"not in my backyard" (NIMBY)** [25] resistance to onshore wind installation, managing fishing rights to open more waters to offshore wind generation, and introducing solar sharing to agricultural land. More renewables would reduce the need for higher-cost technologies such as hydrogen and CCUS, decreasing the average abatement cost of the system by one-third, to $24 per tCO_{2e}.

为了实现这些改变，到 2050 年，日本每年将需要 2 200 万吨氢气，远远高于目前不到 200 万吨的需求。电力部门将消耗 31% 的氢气用于发电。到 2050 年，CCUS 技术需要捕获 1.73 亿吨二氧化碳，这是目前全球碳捕获量的 4 倍以上。为了实现这一目标，日本必须投资 CCUS 技术，建立起庞大的碳储存场站网络，虽然这个网络目前还未有计划。

转型期的资金筹措

依照我们的成本最优路径，日本到 2050 年实现净零排放所需要的总投资为 10 万亿美元，即每年 3300 亿美元。在这些投资中，8 万亿美元将通过重定向资金（redirecting funds）用于对现有技术的投资。其余 2 万亿美元，即平均每年约 700 亿美元，或国家 GDP 的 1%~2%，将用于支付净成本较高的脱碳技术，实现更昂贵的基础设施，如扩大国家电网、纯电动汽车充电站、氢氨传输管道。政府可以利用各种杠杆为转型融资，如补贴、碳税、特定投资或私营部门资金。

替代路径

考虑到有太多的变量在发挥作用，在实施开发成本最优途径的同时，我们还研究了日本可以实现净零排放的一些其他方法。

第一条替代路径，日本可以推动可再生能源发电量占比 80%，而不是 60%。要实现这一目标，需要解决地质和社会方面的制约因素，比如应对邻避效应（NIMBY）对陆地风力发电的抵制，管理捕鱼权以开放更多水域用于海上风力发电，并将太阳能共享引入农业用地。更多的可再生能源将减少对氢和 CCUS 等高成本技术的需求，将系统的平均减排成本降低三分之一至每吨二氧化碳当量 24 美元。

24 In addition to the model using a higher renewables share, we modeled a power-generation mix using a larger contribution of nuclear power. Bringing reactors back online and building an additional 13 GW of capacity (including small modular reactors), **in tandem with** [26] making renewables a higher percentage of the mix, could reduce the average abatement cost 70 percent to $11 per tCO_{2e}. The **energy self-sufficiency rate** [27] of Japan would increase to 88 percent in this scenario, a huge achievement for a country that has always been dependent on fuel imports from abroad. However, relying more heavily on nuclear power is complicated because of the risk of disasters such as Fukushima.

25 In the second alternative pathway, Japan would **outsource** [28] low-value-adding, energy-intensive parts of manufacturing. For example, it could import crude steel from countries with low hydrogen costs and produce the finished steel domestically. To do so would take pressure off Japan's renewables constraints and lower the average cost of abatement for industry from $41 per tCO_{2e} to $34 per tCO_{2e}.

26 However, because this approach would affect only industry, it would reduce total abatement costs by only 8 percent, to $33 per tCO_{2e} compared with $36 per tCO_{2e}, in the central pathway. Implementing this industrial reconfiguration would also require assurances that the outsourced parts of production are fully decarbonized in the producing countries.

　　除了使用更多的可再生能源，我们还模拟了一种发电组合模型，使用更大份额的核能发电。重新启动反应堆，增加 13 吉瓦（1 吉瓦等于 10 亿瓦特）的发电能力（包括小型模块化反应堆），同时提高可再生能源在混合能源中的比例，可以将平均减排成本降低 70% 至每吨二氧化碳当量 11 美元。在这种情况下，日本的能源自给率将提高到 88%，对于一个一直依赖进口燃料的国家来说，这将是一个巨大的成就。然而，大量的核能依赖也是复杂的，因为有发生类似福岛灾难的风险。

　　第二种替代途径，日本将低附加值、能源密集型的制造业进行外包。例如，可以从氢成本较低的国家进口粗钢，然后在国内生产成品钢。这样做将减轻日本可再生能源限制的压力，并将工业减排的平均成本从每吨二氧化碳当量 41 美元降低到 34 美元。

　　然而，由于这种方法只会影响到工业产业，因此总减排成本仅降低 8%，达到每吨二氧化碳当量 33 美元，而在中央路径中则为 36 美元。实施这种工业结构调整还需要保证生产的外包部分在生产国完全脱碳。

　　（文章来源：McKinsey & Company. How Japan Could Reach Carbon Neutrality by 2050. https://www.mckinsey.com/business-functions/sustainability/our-insights/how-japan-could-reach-carbon-neutrality-by-2050. 杨慧、邱馨仪翻译）

词汇 | Vocabulary

[1] densification 致密化，本术语来源于 2017 年联合国人居署的全家专家组会议（EGM）"规划紧凑城市：探索致密化的发展潜力与制约因素"。会议旨在确认应用致密化这一城市转型工具的发展潜力与制约因素，制定优良做法以免产生某些副作用

[2] net-zero target 净零目标。即将所有来源的温室气体排放减少至零，符合《巴黎协定》的目标

[3] societally cost-optimal 社会成本最优

[4] social discount rate 社会贴现率。社会学家将适用于经济活动的贴现率的概念推广到一般社会活动，表明人们对将来发生的各种事情有多重视。一个高的社会贴现率，意味着人们对未来的责任感减弱，只追求眼前利益。社会贴现率受经济贴现率、社会安定等因素的影响

[5] bottom-up analysis 自下而上分析（简称"BU 分析"）。文中意为使用了两个自有模型进行了自下而上的分析

[6] feasibility 可行性。文中意为证明日本净零排放目标的技术可行性，是应用 MPM 分析的目的之一

[7] buildings sector 建筑部门

[8] decarbonize 脱碳。在城市规划领域中，各行业部门进行脱碳，建设"碳中和"城市是重要的目标。在我国，2030 年前实现碳排放达峰，2060 年前实现碳中和的目标也成为 2021 年两会热点话题和"十四五"规划的重要内容

[9] ramp up 爬坡，加大。文中意为扩大纯电动汽车的制造

[10] resort to 诉诸，不得不求助于。文中意为日本在 2030 年后将不得不求助于更昂贵的技术来减少其剩余的碳排放量

[11] abatement cost 减排成本。在我国"双碳"共识下，怎样降低减排成本是重要的研究内容

[12] transition 转型期

[13] primary energy 一次能源，指自然界中以原有形式存在的、未经加工转换的能量资源，又称"天然能源"

[14] renewable 可再生能源

[15] clean hydrogen and ammonia 清洁的氢氨能源

[16] threefold 三倍

[17] would have to 不得不

[18] infrastructure 基础设施

[19] reinforce 加强，强化

[20] redirecting fund 重定向投资

[21] incumbent technology 现有技术

[22] finance 文中意为融资，指利用各种杠杆为转型融资。

[23] subsidy 补贴

[24] private-sector funding 私营部门资金

[25] not in my backyard (NIMBY) 邻避效应，指居民或当地单位因担心建设项目的负面影响而激发的嫌恶情结。文中意为需要应对当地居民和单位的邻避效应对陆地风力发电的抵制

[26] in tandem with 同时

[27] energy self-sufficiency rate 能源自给率

[28] outsource 外包，文中意为将低附加值、能源密集型的制造业进行外包

练习与思考 | Comprehension Exercise

1. 请根据句中的关键词，查阅相关资料，深入了解，并分组讨论。

（1）**Battery manufacturers** would also need to shift production to meet the new demand, which would increase eightfold in the next decade.

（2）The government can utilize various levers to finance the transition, such as **subsidies, carbon taxes, special investments, or private-sector funding.**

2. 根据本段内容，讨论 NIMBY 的影响因素及应对策略。

Achieving this would require addressing geological and social constraints, such as confronting "**not in my backyard**" **(NIMBY)** resistance to onshore wind installation, managing fishing rights to open more waters to offshore wind generation, and introducing solar sharing to agricultural land.

课后延伸 | Reading Material

基础设施的碳成本——气候危机的关键 | Carbon Cost in Infrastructure—the Key to the Climate Crisis

Carbon Cost in Infrastructure—the Key to the Climate Crisis

1　To prevent catastrophic climate change, we must tackle the carbon emissions associated with national infrastructure. Globally significant quantities of carbon are emitted through its construction; however, these emissions are often overlooked in favour of carbon from its use. To understand the whole picture we must consider whole life carbon, just as we consider whole life cost.

Carbon, Cost and the Tipping Point

2　There is a close relationship between carbon and cost in the infrastructure sector. In most cases the relationship is approximately proportionate—if carbon is reduced, so is cost. There are therefore both environmental and financial incentives to ensure carbon emissions are minimised.

3　Although "reduce carbon, reduce cost" is a simple general principle in infrastructure, there are many instances where the lowest carbon solution is not the least expensive. There are carbon cost tipping points that need to be assessed to provide the most worthwhile carbon reduction. It has become clear that if governments intend to reach net zero carbon emissions, detailed assessment of these tipping points is required.

Managing Infrastructure Carbon

4　The minimisation of carbon emissions in infrastructure is not only an environmental discipline—it is of much wider relevance when our overarching goal is sustainable development. To most effectively reduce carbon, we need to integrate carbon and cost reduction across all design disciplines from the outset and at the highest level. There is a necessary evolution for carbon reduction in infrastructure: from carbon counting, to carbon management, to carbon cost management. It is only through carbon cost management that we can minimise carbon as efficiently as possible.

5　To be most effective, carbon cost management must be integrated into the design process. From the outset and at every decision point, carbon emissions and associated cost should be considered. Carbon must become a natural part of the value engineering process and be considered alongside cost to enable best-value reductions.

基础设施的碳成本——气候危机的关键

为了阻止灾难性的气候变化，我们必须解决与国家基础设施相关的碳排放问题。在全球范围内，大量的碳排放来自基础设施建设；然而，这些排放往往易被忽视。为了掌握碳排放的全貌，我们必须考虑全生命周期的碳排放，正如我们考虑全生命周期的成本一样。

碳排放、成本和临界点

基础设施领域的碳排放和成本之间存在密切的关系。在大多数情况下，这种关系是近似成比例的——如果碳排放减少，成本也会减少。因此，可以同时实施环境措施和财政激励，确保将碳排放降至最低。

尽管"减碳排、降成本"在基础设施建设中是一个简单的原则，但在许多情况下，最低碳的解决方案并非是最便宜的。因此，需要评估碳排放成本临界点，以提供最佳价值的碳减排。很明显，如果政府打算实现净零碳排放，就需要对这些临界点进行详细评估。

管理基础设施的碳排放

基础设施碳排放的最小化不仅是一项环境学科——当我们的总体目标是可持续发展时，它具有更广泛的意义。为了最高效地减少碳排放，我们需要在一开始就从总体上整合所有设计环节中的碳排放和成本：从碳排放核算到碳排放管理，再到碳排放成本管理，这是基础设施碳减排的必然演进。只有通过碳排放成本管理，我们才能尽可能高效地减少碳排放。

为了实现最佳效果，碳排放成本管理必须被纳入设计过程。从一开始，每个决策点的制定，都应该考虑碳排放及其相关成本。碳排放必须成为价值工程流程的一个自然组成部分，并与成本一起考虑，以实现最佳价值的碳减排。

6   Carbon cost management can be mandated at a strategic level through the application of a specification such as *PAS 2080*. This is internationally relevant and ensures a consistent focus on carbon and cost reduction.

Making Carbon Reduction Contractual

7   The procurement process is key to carbon cost management. From the inception of a project and at each procurement stage, carbon reduction should be a differentiating metric. Innovations in procurement can incentivise carbon reduction and accelerate a variety of low-carbon solutions. Carbon reduction in the procurement process should not only be qualitative input to a sustainability assessment – it should be a fundamental and measurable contractual requirement.

Leadership

8   Without strong leadership, carbon reduction is, at best, a passive environmental aspect that has no bearing on a design. With strong leadership, proactive carbon cost management occurs at every design decision point. The client is aware of all carbon reduction opportunities: which will save money, which will cost money, which are affordable, which are most worthwhile. Carbon and cost are presented on the same page.

9   While the focus of this report has been on infrastructure carbon cost management, this assessment approach is equally applicable to every aspect of the built environment. In fact, carbon cost assessment is applicable to decisions we all face in our daily lives, from the best way to visit relatives over the holidays to your choice of meal this evening.

10 Everything with a financial cost has an associated carbon value. Understanding these carbon cost relationships gives us the knowledge to help us most efficiently reduce carbon and prevent climate change.

11 We all have a role to play in reducing carbon emissions, but the greatest impact requires decisive leadership at all levels.

碳排放成本管理可以通过应用《PAS 2080》等规范在战略层面强制实施。这不仅具有国际意义，也保证了对减少碳排放和成本的持续重视。

让碳减排成为一种合约

采购过程是碳排放成本管理的关键。从项目初始到采购的各个阶段，碳减排都应该是考虑的指标。采购方面的创新可以激励减少碳排放，并加速各种低碳解决方案的实施。在采购过程中，碳减排不仅应该作为可持续评估的定性考量，同时还应成为一项基本的、可衡量的合同要求。

领导层

如果没有强有力的领导层，碳减排充其量只是一种被动的行为。在强有力的领导下，在每个设计环节和决策点我们都可以进行主动的碳排放成本管理。让客户了解所有的碳减排机会：哪些可以省钱，哪些需要花钱，哪些是可负担的，哪些是最值得的，并同时了解碳排放和成本的信息。

虽然本报告的重点是基础设施的碳排放成本管理，但该评估方法同样适用于建成环境的各个方面。同样地，碳排放成本评估也适用于我们日常生活中面临的抉择，比如从节假日走亲访友的最佳出行方式到今晚的用餐选择。

事实上，任何有财务成本的东西都有其相应的碳价值。了解这些碳排放与成本的关系可以帮助我们最高效地减少碳排放，并遏制气候变化。

我们都可以在减少碳排放方面发挥作用，但要想发挥最大的影响力，则需要各层级的果断领导力。

POLICY MAKERS enable the appropriate market framework and regulations to encourage carbon cost management.

CLIENTS recognise your role in establishing carbon reduction as a key contractual metric across your supply chain.

ENGINEERS consider both the carbon and cost impacts of your decisions and present these together.

CONTRACTORS promote low-carbon innovations either side of the carbon cost tipping point.

EDUCATORS prepare the next generation of carbon cost experts through science, technology, engineering and mathematics.

政策制定者——建立适宜的市场框架和规范,以鼓励碳排放成本管理。

客户——考虑整个供应链中的碳减排,确保合同中包含碳减排相关的关键指标。

工程师——考虑碳减排和成本对你的决定的影响,并把它们一并提出。

承包商——推动碳排放成本临界点两侧的低碳创新。

教育工作者——通过科学、技术、工程和数学教学,培养下一代碳排放成本专家。

（文章来源：SWECO. Urban Insight Report—Carbon Cost in Infrastructure—the Key to the Climate Crisis.https://www.swecourbaninsight.com/climate-action/carbon-cost-in-infrastructure-the-key-to-the-climate-crisis. 张鹤鸣、王琛芳、张浩然、赵柄智、冷怡霖、张佳玥、陈俊豪、杨蕊源、林煜翻译）

第 4 课　城市交通

lesson 4　Urban Transport

导读 | Introduction

全球城市应对交通新挑战 | Global Cities Tackling New Transportation Challenges

　　本单元课文选自麦肯锡 2021 年 7 月发布的 *Elements of Success: Urban Transportation Systems of 25 Global Cities*（《成功的要素——全球 25 城交通报告》），这是继 2018 年之后该主题的第二次报告。2018 年至今，全世界的城市都在开展大规模的建设，意图改善交通系统的运行状况。城市遇到了新的挑战，包括环境安全、移动性的改变等。由于新冠肺炎疫情，2020 年成为了一次大考，极大影响了各大城市交通系统的运行，该报告即是对最近的交通变化展开的一次新研究。

预备知识 ｜ Preliminary Knowledge

样本城市的选择依据与可用性指标评级 ｜ Selection of Target Cities and Ratings Based on Availability Metrics

研究通常分为 5 个阶段进行：选择样本城市、制定指标计算评估值、确定指标权重并编制评级表、增加附加评估以及与城市居民的主观感知进行对比。其中，样本城市的选择依据如下。

城市人口应超过 500 万，并在全国经济中起主导作用。人均本地生产总值（GRP）超过 1 万美元，汽车数量超过每千人 150 辆，数据来源应能在国际数据源中公开查询。

该报告在 13 800 个城市中筛选出 21 个拥有相似交通系统的城市，并另外增加 4 个具有研究相关性的城市（上海、新加坡、柏林和香港）进行详细阐述。

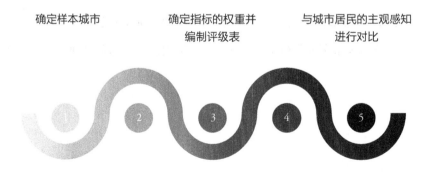

	城市数量	选择标准
1	13 800	城市体量 城市是全国最具经济意义的城市之一
2	38	经济发展水平 人均本地生产总值至少 1 万美元
3	32	交通系统指标 每 1 000 名居民拥有 150 辆汽车
4	28	数据可用性和质量 > 50% 的数据来源于国际数据源
5	21	专家评估 在 ≥ 2 个评级中处于领先地位，人口为 > 300 万

4 个附加城市

可用性指标评级

可用性指数是由一组用于评估城市居民可用出行方式的指标计算得出的。这些指标主要包括轨道交通、道路交通、共享交通和外部连通性等四个维度。

这一指数描述了城市在轨道交通、道路交通、共享交通以及与其他城市经由航空目的地连通等方面的可用性。

可用性指数最高的三座城市分别为伦敦、巴黎以及马德里。英国首都伦敦拥有最多数量通航交通的城市：在新冠肺炎疫情之前，希斯罗机场拥有超过 450 个国内和国际航班。同时，伦敦在城市道路交通网络可用性方面也是世界领先。例如，自行车道长度在过去三年间增长了 30% 以上，自行车道里程在总道路里程中占据越来越大的份额。

巴黎仅仅因为在外部连通性上的得分低于伦敦，而在排名中屈居第二。巴黎在步行基础设施的衔接方面（即次级道路交通网络的建设）领先于伦敦，意味着巴黎拥有多分支网络状的道路交通，即使不沿直线行进，从 A 点步行到 B 点也不会浪费太多时间。

排名第三的马德里，在汽车共享指标上表现出色（每百万人汽车保有量为 840 辆）。除此之外，这座城市还拥有发达的轨道交通网络，在这方面只有东京拥有微弱的优势。在这两座城市中，有大约 91% 的人口都居住在距离地铁和通勤铁路车站步行 20 分钟的范围内。

就可用性而言，城市的理想形态应当是伦敦（外部连通性）、东京（轨道交通网络）、米兰（道路交通网络）和北京（自行车数量以及可用于共享交通的机动车数量）的结合体。

课文讲解 | Text Explanation

Elements of Success: Urban Transportation Systems of 25 Global Cities 2021 (section)

1. Moscow Central Diameters[1] (MCDs) (Public-transport infrastructure development)

The Ivolga new-generation train has 11 cars and carries more than 3,000 passengers.

Project description

1 The Moscow Central Diameters join fragmented railroad directions, creating fully functional lines that can be used to cross Moscow nonstop and reach the nearest neighboring cities. Railroad lines connecting Moscow and its **immediate environs**[2] comprise a **single**[3] transport system including the Moscow underground. The first two diameters were opened in 2019.

Uniqueness of the project

2 Construction of the MCDs has become one of the world's largest **city rail transport development projects**[4]. After their launch, the MCDs have greatly expanded the underground-integrated system, improving quality of life for millions of **Muscovites**[5].

成功的要素：全球 25 城交通系统报告（节选）

1. 莫斯科中央直径线（公共交通基础设施建设）

伊沃尔加新一代列车，可以加挂 11 节车厢，搭载 3000 多名乘客。

项目概况

莫斯科中央直径线是连接城市及区域零散铁路线路，支撑打造功能齐全、一体化网络的轨道交通线路。它可以穿越莫斯科到达最邻近的城市，由此连接了莫斯科及其周边地区的铁路线，构成了包括莫斯科地铁在内的统一运输系统。2019 年，莫斯科开通了最早两条中央直径线。

项目特点

莫斯科中央直径线是目前世界上最大的城市轨道交通建设项目之一，自它开通运营以来，极大提升了城市地下轨道系统一体化、集成运行能力，提高了数百万莫斯科居民的生活水平。

3   The new diameters are used by new-generation Ivolga (Oriole) trains, each with 11 cars, that can carry more than 3,000 passengers. The trains feature numerous functions modern city residents may need, including free Wi-Fi, USB gadget chargers, and bicycle racks.

Project impact [6]

4   Although the COVID-19 pandemic has driven down mobility, the project has already produced positive results. In particular, it has increased **accessibility** [7] of the underground-integrated network, improved transport system performance, saved passengers considerable time, and reduced traffic at certain underground lines by **redistributing** [8] passenger flows away from the most heavily used **line sections** [9].

5   •38%—Moscow survey respondents who mentioned the project

•132 km—Total length of MCDs

•60 Number of stations

•20%—Increase in length of the underground network

•22%—Increase in number of stations in the system integrated with the underground

•130 million—MCD passenger traffic per year

•8%—Share of trips in the system integrated with the underground

莫斯科中央直径线采用了新一代的 Oriole 车型，拥有 11 组车厢，能搭载 3000 多名乘客。车厢内配备有 Wi-Fi、USB 充电接口、自行车架等功能设施，以满足现在城市居民所需。

项目成效

尽管新冠肺炎疫情降低了城市出行的流动性，但莫斯科中央直径线项目还是产生了积极的效果。尤其是它有效提升了城市地下交通系统的可达性，提高了轨道交通服务性能，节省了乘客出行时间，同时通过客流量重分配缓解了某些地下线路高峰时期重点断面的客流压力。

- 莫斯科城市调查中提及该项目的人数占比：38%

- 线路总里程：132 千米

- 站点数量：60 个

- 地下轨道交通网络里程提升比例：20%

- 与地下交通系统连通的站点增加比例：22%

- 每年使用人次：1.3 亿

- 与地下交通系统联运比例：8%

2. Hong Kong's Northern Connection: Tuen Mun-Chek Lap Kok Link (TM-CLKL) (Road infrastructure)

The Tuen Mun-Chek Lap Kok undersea tunnel has a depth of 50 meters and length of 5 kilometers; no fare is charged.

Project description

1   The Tuen Mun-Chek Lap Kok **undersea tunnel** [10] was opened in Hong Kong in 2020. Together with the Southern Connection, it will form a strategic route joining the Northwest New Territories with the Hong Kong-Zhuhai-Macau Bridge, Hong Kong Port, and Hong Kong International Airport/Northern Lantau. Tuen Mun-Chek Lap Kok is the longest and deepest road tunnel in Hong Kong, with a diameter comparable to the height of a six-story building.

Uniqueness of the project

2   TM-CLKL is an innovative undersea tunnel construction project. Several records were set during its **implementation** [11]. In particular, the builders employed the world's largest tunnel-boring machine (with a diameter of 17.6 meters) to bore, using the shield method, the deepest (50 meters) and longest (5 kilometers) tunnel in Hong Kong.

3   Special attention was paid to environmental protection. For example, more than 280 **insectivorous** [12] plants **on the brink of extinction** [13] were moved to alternative locations. Also, the **boring method** [14] was selected, at least in part, because of its lesser impact on the water environment.

2. 香港北部屯门—赤鱲角干线（道路基础设施建设）

屯门—赤鱲角干线海底隧道深 50 米、长 5 千米，免费通行。

项目概况

香港屯门—赤鱲角海底隧道在 2020 年开通运行，北部干线与南部干线共同构成连接新界西北地区与港珠澳大桥、香港港口、香港国际机场、北大屿山的战略道路网络。这条海底隧道是目前中国香港地区最长、最深的地下隧道，高度约等于 6 层楼高。

项目特点

该项目是海底隧道建设的一次创新，在施工过程中创下多项记录：采用目前世界上最大的隧道掘进机（直径 17.6 米）进行盾构法施工、是香港最深（50 米）且最长（5 千米）的隧道。

同时，该项目很好地践行了生态友好理念，例如采用对水环境影响较小的隧道开挖技术、施工中将濒临灭绝的 280 种食虫植物移植到合适的地方。

Project impact

4 Construction of the tunnel greatly improved **connectivity**[15] between the southern part of Tuen Mun County and the Hong Kong International Airport.

5 •33%—Hong Kong survey respondents who mentioned the project

•5.5 km—Total length of tunnel

•22 km—Decrease in length of trip to the airport for Tuen Mun County residents

•20 min—Decrease in **duration**[16] of trip to the airport for Tuen Mun County residents

3. Display of Bus Load Data in New York's MYmta Application (Transport systems digitization)

In New York, 550,000 people use the application every week.

项目成效

项目极大地提升了屯门南部地区与香港国际机场的连通性。

· 香港城市调查中提及该项目的人数占比：33%

· 隧道总里程：5.5 千米

· 屯门地区居民去机场距离减少：22 千米

· 屯门地区居民去机场时间减少：20 分钟

3. 纽约 MYmta 应用展示公交载客量（交通系统数字化）

在纽约，每周有 55 万人使用应用程序辅助出行。

Project description

1 In New York City, MTA passengers use the MYmta application to plan their trips through the regional transit network [17]. The planning process takes into account services offered by other transport organizations, such as the Staten Island Ferry, NYC Ferry Service, PATH, and NJ Transit.

2 The MYmta application was updated in July 2020 to reflect COVID-19 realities. In particular, it became possible to track the number of passengers on arriving trains and buses. This enables passengers to plan their trip routes so they **comply with social-distancing rules** [18].

Uniqueness of the project

3 Such a function had never before been launched in the United States on a full-scale basis. Special infrared sensors and 3-D object-recognition technologies are used to determine the number of passengers. The sensors are placed above bus doors and connected to GPS trackers installed on each vehicle.

Project impact

4 Using the application, passengers can quickly assess how full the approaching vehicle is. This gives them an opportunity to do early route planning so they are more likely to mitigate the risk of infection and comply with social distancing rules.

5 •10%—New York survey respondents who mentioned the project

   •40%—Share of vehicles with onboard sensors

   •550,000—Passengers using the feature every week

项目概况

纽约大都会运输署（纽约公交车运营商）的应用程序 MYmta 可用于规划区域交通出行方案。该应用提供了不同公交服务运营商的数据，比如 "Staten Island Ferry" "NYC Ferry Service" "PATH" 和 "NJ Transit"。

2020 年 7 月，为防控新冠肺炎疫情，MYmta 更新了系统功能，可以即时查询公交车上的乘客数量，辅助乘客制订出行规划，以此保证遵守社交距离规定。

项目特点

该功能之前从未在美国大范围推出，它采用了先进的感应设备与 3D 物体识别技术来确定公交车上的乘客数量。感应设备通常装在车门上，并与 GPS 追踪器连接。

项目成效

通过该应用，乘客可以很快知道公交车上的搭载情况，并依此提早规划出行路线，减少不必要的社交接触，降低潜在病毒感染风险。

• 纽约城市调查中提及该项目的人数占比：10%

• 车辆感应器安置率：40%

• 每周使用乘客数量：55 万人

4. Milan's Strade Aperte Cycling and Pedestrian Infrastructure (Cycling and pedestrian infrastructure)

Creation of bicycling infrastructure in Milan aims to increase the safety of all residents.

Project description

1 Milan has conducted an experiment involving rapid expansion of pedestrian and cycling space to protect city residents in the wake of [19] COVID-19. (The purpose of the Strade Aperte (Open Streets [20]) project is to shape a new approach to mobility and public spaces, making the city more environmentally friendly and comfortable.) The project, announced in 2020, envisages reprofiling of 35 kilometers of streets and expansion of the existing speed limit—30 kilometers per hour—to new areas across the city.

Uniqueness of the project

2 The project has drawn praise from the expert community as an example of a new perspective on street design. It is intended not only to ensure that cars can move from point A to point B in the shortest possible time, but also to increase the safety of all residents moving across the city.

3 The project's purpose is to give Milan residents protected and accessible streets, create new public spaces, and encourage walking and riding bikes or scooters [21] as alternatives to public transport and personal cars.

4. 米兰"开路计划"（行人与自行车设施建设）

米兰自行车基础设施的建设旨在提高所有居民的安全。

项目概况

为应对疫情影响下出行方式的变化，米兰实施了扩大行人与自行车活动空间的一系列举措。"开路计划"的目标是实施新的交通组织、塑造新的公共空间，使城市环境更加友好，居民出行更加舒适。"开路计划"在 2020 年提出，旨在重塑城市道路公共活动空间，提出重新设计全市 35 千米的城市道路，拓展城市车辆限速 30 km/h 的范围。

项目特点

该项目不光是为了提升小汽车的通行效率，更多是为了提高所有城市居民的交通出行安全性，这种从新视角进行街道设计的做法也被当作范本，并获得了专家的赞赏。

该项目旨在为米兰市民打造安全、可达的道路及焕然一新的公共活动空间，并鼓励以步行与骑行方式替代私家车和公共交通工具。

Project impact

4   The key purpose of the project is to maintain a transport balance in the city—a balance that can be disrupted if residents make excessive use of personal cars. City authorities hope to prevent a resurgence of private car use as city residents return to their offices avoiding overcrowded public transport.

5   •18%—Milan survey respondents who mentioned the project

   •35 km—Total length of reprofiled streets

   •60%—Target share of streets where a 30 km/h speed limit is in effect

项目成效

该项目的主要目的是维持城市交通平衡——如果居民过度使用私家车就会打破这种平衡。居民因为不愿意乘坐拥挤的公共交通工具而选择私家车出行，城市当局希望减少私家车使用，避免私家车交通增长回潮。

- 米兰城市调查中提及该项目的人数占比：18%

- 改造道路总里程：35 千米

- 目标限速 30 km/h 道路占比：60%

5. Seoul's "Smart Shelter" [22] Bus Stops (Travel comfort)

Project description

1   "Smart shelters" are designed to protect people from summer heat and monsoon rains and to combat the spread of COVID-19. They are glass cubes equipped with an air conditioner and a UV sterilizer [23] to ventilate [24] and cool air. In addition, the shelters are fitted with surveillance cameras [25] and digital screens to warn passengers that a bus is approaching.

2   They emerged in Seoul streets in August 2000. Each stop has hand sanitizers [26], free Wi-Fi, and plugs that can be used to charge mobile devices or notebooks. Heat visualization cameras [27] are used to let in passengers only if their body temperature does not exceed 37.5 degrees Celsius.

Uniqueness of the project

3   New bus stops share real-time information with police and fire departments. To do that, they use smart video surveillance cameras, alarm signals, and smart noise sensors, thus enabling those services to minimize their response times.

4   Median barriers [28] are replaced with green plants, giving new transport stops a city-garden appearance [29]. Passengers are offered wireless mobile-phone chargers, air cleaners, and free Wi-Fi.

Project impact

5   During the week after installation, smart stops were used by 300 to 400 people per day. The Seoul municipal authority expects that introduction of smart shelters will not only improve the quality of services provided by the city transport system, but also reduce social security costs [30] related to mitigation of harm caused by small dust particles and expand application of smart technologies across the city.

6   • 10%—Seoul survey respondents who mentioned the project

      • 37.5°C—Maximum permitted body temperature of stop users

      • $84,000—Cost of a bus-stop installation

5. 首尔智能候车亭（舒适出行）

项目概况

"智能候车亭"的设计旨在保护人们免受高温和暴风雨的影响，同时抗击新冠肺炎疫情的传播。这些巴士站外观为带玻璃墙的小盒子，内部装有空调和紫外线消毒器，用于通风和提供冷气。此外，候车亭还安装了监控摄像头和数字屏幕，提醒乘客公交车即将到达。

这些候车亭自 2000 年 8 月出现在韩国首尔的街头。每个站点都配有洗手液、免费 Wi-Fi 和为手机或笔记本电脑准备的充电口。热成像摄像头只允许体温不超过 37.5 摄氏度的乘客进入候车亭。

项目特点

这些新的巴士站和公安及消防部门共享实时信息。通过采用智能视频监控摄像头、警报信号和智能噪声传感器等科技手段减少服务响应时间。

护栏被绿色植物取代，为新的站点营造城市花园景观。乘客可以享受无线手机充电器、空气净化器和免费上网服务。

项目成效

在新候车亭启用后的一周内，每天有 300 至 400 人使用。首尔市政府认为，应用这些智能候车亭不仅可以提高城市交通系统的服务质量，还可以减少微小颗粒物等空气污染带来的治理费用，并扩大智能技术在全市范围的应用。

- 首尔城市调查中提及本项目的受访者占比：10%
- 候车亭使用者体温不可超过：37.5 ℃
- 每个候车亭的安装费用：84 000 美元

6. London's Wood Lane Arches [31] (Transit-oriented Development)

Shops, galleries, and public spaces have been created in an area adjacent to the Wood Lane tube station.

Project description

1   A series of railway arches next to the Wood Lane tube station are to be converted into shops, galleries, and community spaces in a project **backed by** [32] Hammersmith and Fulham Council.

2   Of the 19 arches in the planned first phase, 13 will be used for retailing, while the other six will provide new pedestrian routes, bicycle parking, and storage facilities.

3   The arches area will be opened for pedestrians, with vehicular traffic to be banned; however, a parking lot is available in the nearby Westfield London mall. Location of the arches offers convenient access to London's bicycle route network. The project envisages 66 safe sections for cyclists, both personnel and visitors.

Uniqueness of the project

4   Completion of the project is intended to enable the transformation of unused transport infrastructure spaces. That will affect not only retail facilities, but also city residents who use the new pedestrian and biking routes.

6. 伦敦 Wood Lane 拱门（公共交通导向发展）

Wood Lane 地铁站附近的区域增加了新的商店、画廊和公共空间。

项目概况

伦敦 Wood Lane 地铁站旁边的一系列拱门将被改造成商店、画廊和社区空间。本项目由 Hammersmith and Fulham 区议会支持改造。

项目第一期包含 19 个拱门，其中 13 个将用于零售用途，而其他 6 个将作为人行道、自行车停放处和储存服务处。

拱门区域将对行人开放，禁止车辆通行，但附近的 Westfield London 购物中心含有停车场。拱门所在的位置可以便捷地汇入伦敦的自行车道路网。本项目为骑行者（工作人员和游客）设计了 66 个安全区域。

项目特点

本项目旨在改造未被利用的交通基础设施空间。此类改造不仅可以提供零售场所，还可以为居民增加人行道和自行车道。

Project impact

5   The White City renovation area [33] covers five key facilities, which form an impressive group of top-quality residential, office, retail, and community spaces. When project implementation started, the arches under the Circle line and the Hammersmith and City line (built in the 1860s) were not open for the public; they were filled with debris, loose stone, and waste. Upon completion of the project, the area will get additional retail space (2.3 million square feet), office space (2.2 million square feet), and 5,000 new houses.

6   •6%—London survey respondents who mentioned the project

      •66—Sections reserved for bikers in the Wood Lane project

      •2.3 million sq. ft.—New retail space

      •2.2 million sq. ft.—New office space

      •5,000—Residential units to be built under the project

项目成效

White City 城市更新区包含五个关键设施，形成了一系列引人注目的高端住宅、办公、零售和社区空间。当本项目开始实施时，伦敦地铁的环状线及 "Hammersmith and City" 线（建于 19 世纪 60 年代）下的拱门不对公众开放，且拱门里充满碎片、石头和杂物。本项目改造后，这片区域将增加 230 万平方英尺（1 平方英尺 ≈ 0.092 9 平方米）零售面积、220 万平方英尺办公面积和 5 000 套居住单元。

· 伦敦城市调查中提及本项目的受访者占比：6%
· 本项目为骑行者预留的安全区域：66 个
· 本项目改造后增加的零售面积：230 万平方英尺
· 本项目改造后增加的办公面积：220 万平方英尺
· 本项目计划建造的居住单元数量：5 000

（文献来源：MaKinsey & Company. Elements of Success: Urban Transportation Systems of 25 Global cities. https://www.mckinsey.com/business-functions/operations/our-insights/building-a-transport-system-that-works-five-insights-from-our-25-city-report. 李曼婷、李莹、林煜、雷雪飞、郑思琦、成嘉琪、余铭航、苗雨田、周航、鲁雨萌、张佳玥、杨慧、郭文韬、谢瑾、尤雨婷、郭玥、东方、仲璨文、赵柄智翻译）

词汇 | Vocabulary

[1] central diameters 中央直径线

[2] immediate environs 紧邻的周边环境

[3] single 单一的，文中意为铁路线和地铁线统一为一套运输系统

[4] city rail transport development project 城市轨道交通建设项目

[5] Muscovite 莫斯科居民

[6] impact 成效

[7] accessibility 可达性，文中意为提升了城市地下交通系统的运行服务能力

[8] redistribute 重分配，文中意为客流的流量重分配缓解了高峰时期重点断面的客流压力

[9] line section 道路断面

[10] undersea tunnel 海底隧道

[11] implementation 实施，施工过程

[12] insectivorous 食虫的

[13] on the brink of extinction 濒临灭绝

[14] boring method 开挖技术

[15] connectivity 连通性

[16] duration 时耗

[17] transit network （公交）运输网络

[18] comply with social-distancing rules 遵守社交距离规定

[19] in the wake of 在……之后

[20] Open Streets 开路计划。意大利语为 "Strade Aperte"，是基于疫情封锁而产生的想法，鼓励骑自行车穿过街道

[21] scooter 滑板车。文中意为和步行、骑自行车都是慢行交通的方式

[22] smart shelter 智能候车亭

[23] UV sterilizer 紫外线消毒器

[24] ventilate 使通风，使通气

[25] surveillance camera 监控摄像头

[26] hand sanitizer 洗手液

[27] heat visualization camera 热成像摄像头

[28] median barrier 护栏

[29] city-garden appearance 城市花园景观

[30] social security cost 社会保障成本

[31] arch 拱门

[32] be backed by 被给予支持

[33] renovation area 更新区域

练习与思考 | Comprehension Exercise

1. 请根据句中关键词，查阅相关资料，深入了解，并分组讨论。

（1）Milan has conducted an experiment involving rapid expansion of pedestrian and cycling space to protect city residents in the wake of COVID-19. The purpose of the **Strade Aperte (Open Streets)** project is to shape a new approach to mobility and public spaces, making the city more environmentally friendly and comfortable.

（2）Construction of the MCDs has become one of the world's largest **city rail transport development projects**. After their launch, the MCDs have greatly expanded the underground-integrated system, improving quality of life for millions of Muscovites.

课后延伸 | Reading Material

"慢街"政策与城市规划 | Slow Streets' Disrupted City Planning

Slow Streets' Disrupted City Planning

1   "I think there's a tension between planners wanting to act fast, because their work is so critical to reduce fatalities and greenhouse gas emissions—the reasons for this work are so compelling and historic," said Corinne Kisner, the executive director of the National Association of City Transportation Officials. "But the urgency to move fast is in conflict with the speed of trust, and the pace that actually allows for input from everyone who's affected by these decisions."

In May, the Los Angeles Department of Transportation (LADOT) closed streets in West Los Angeles to limit traffic and promote social distancing.
(Source: CHRIS DELMAS/AFP via Getty Images)

2   Indeed, while some cities took a lashing for equity missteps and oversights, those that hadn't acted swiftly to shut down streets also faced heavy criticism from bicycle and pedestrian advocates for missing their chance to reclaim auto-dominated streets, as in Chicago. But slow adoption could also be a virtue. After initially deciding against a slow streets program without rigorous community outreach, the city of Atlanta later set up Covid testing sites and census registration tents, programmed by trusted community groups, along key neighborhood corridors.

3   In Los Angeles, demand for slow streets came early on in the pandemic from wealthy, white neighborhoods, where residents were working from home. But rather than react with a citywide rollout, LADOT general manager Seleta Reynolds took a different tack. Low-income communities of color were "still on transit,"

"慢街"政策与城市规划

美国全国城市交通官员协会（National Association of City Transportation Officials）的执行主任科琳·基斯纳（Corinne Kisner）认为，慢街政策中体现了与城市规划的矛盾，规划师想要尽快采取行动，因为他们的工作对减少交通死亡、减少温室气体排放至关重要——开展这项工作的理由非常充分，而且具有历史意义。但是，行动的紧迫性与人们建立信任的速度相冲突，建立信任的速度实际上就是允许受决策影响的每个人提供他们意见的速度。

5 月，洛杉矶运输部（LADOT）封锁了西洛杉矶的街道，以限制交通并推广社交距离。
（图片来源：CHRIS DELMAS/AFP via Getty Images）

的确，一些城市由于封锁街道时存在社会公平和种族平等方面的疏漏而备受争议，但那些没有迅速采取行动关闭街道的城市也同样面临着来自非机动车和行人立场的批评，谴责它们错失了改变机动车街道统治地位的机会，就像我们在芝加哥看到的那样。但缓慢推进也可能是更好的选择。亚特兰大市在没有社区帮助的情况下决定暂缓推行慢行街道。随后，该市在主要的社区道路旁设立了新冠病毒检测点和人口普查登记帐篷，委托当地社区组织负责。

在洛杉矶，对慢行街道的需求最初源自富裕的、居家工作比例很高的白人社区。但 LADOT 的总经理 Seleta Reynolds 并未立刻在全城范围内推广慢行街道。她在最近的一次研讨会上表示，低收入的有色人种社区目前仍处于过渡阶段，因为他们更常开车。当低收入人群失去了零售或服

she said on a recent panel. "They were the ones who were actually driving more because as they lost their jobs in retail and service industries, they fell into jobs in the gig economy where driving itself was a job. And they needed different things and were being disproportionately impacted by the virus."

4   So the city paid community-based organizations to survey hard-hit neighborhoods about what they wanted, Reynolds said, then used funding earmarked for slow streets to support outdoor dining in those areas. Other neighborhoods that wanted street closures still received them.

5   Durham also found success by listening and responding. The transportation department applied for and won a $25,000 grant from NACTO to implement a transportation-based pandemic response in partnership with a community-based organization. Their program was dubbed Shared Streets, and in East Durham, it kicked off with three public meetings led by Ortiz. Through those conversations, longstanding concerns about traffic speeds and safety ultimately influenced the design of the program. In the end, five streets received curb extensions and painted traffic circles to slow speeding cars, created with the help of neighborhood volunteers.

Aidil Ortiz speaks to residents in East Durham about the Shared Streets program.
(Source: Courtesy City of Durham)

务业的工作后，他们不得不靠打零工为生，而开车本身就是一种工作。不同的群体受疫情影响的程度是不同的，因此他们的需求也并不相同。

Reynolds 说，市政府拨款给社区组织，让他们调查受新冠影响最严重的社区，了解他们的需求，然后用慢行街道的专项资金来扶持这些地区的室外餐饮业。与此同时，其他社区仍然会得到此笔款项，用于慢行街道的建设。

达勒姆的疫情应对计划也取得了成功。该市交通部门申请并获得了来自 NACTO 的 25,000 美元捐款，用于与社区组织开展合作，实施基于交通运输的疫情应对措施。他们的计划被命名为共享街道。Ortiz 在东达勒姆组织召开了三次公开会议，以宣布计划开启。对车速和道路安全的长期关注最终得以落地设计。最终，在社区志愿者的帮助下，五条街道得到了拓宽，并增设了路中环岛以减缓车速。

Aidil Ortiz 在东达勒姆向居民介绍共享街道计划。

（图片来源：Courtesy City of Durham）

6   "I think taking the extra time did allow us to develop a project that I hope is better serving the needs of communities," said Dale McKeel, the city's bicycle and pedestrian coordinator. He credits the city's relationship with Ortiz and its existing commitments to equitable engagement.

7   To Ortiz, the program was essentially a small-scale trust-building exercise between the city and constituents who harbor legitimate suspicions. The fact that the shared streets project was limited in scale and temporary helped, she thinks. "It was reassuring to people that this wasn't just a one-time thing that we would listen to them just this once," she said. "We are committed to listening."

8   Now, as officials consider extending these programs while grappling with pandemic-battered budgets, the challenge is to keep communication lines open. Reaching out to residents, holding genuine conversations and incorporating feedback requires staff time as well as money. Building trust with marginalized communities may also require cities to first put their trust in key intermediaries. Tamika Butler and Naomi Iwasaki, transportation consultants who focus on social justice and who advised the NACTO grant program, said that a central lesson from Durham, Atlanta and L.A. is that officials allowed the needs of underserved people to determine solutions on the ground, even if it meant expanding how problems were originally defined.

9   "Having community partners take the lead—which is often a challenge because of built-in bureaucracies—and being open to hearing stuff that is not always transportation-related is a huge part of it," Butler said. "Cities need to understand that no one issue lives in isolation."

City planning's "history of trauma"

10 Slow Streets was more than a program—it also became a turning point in urbanist discourse. Several equity advocates interviewed for this article said they believed that the planners and officials they work with have developed a deeper understanding of the issues at stake as a result.

11 Thomas, who wrote the op-ed that critiqued Oakland's program, says she's still disappointed by outcomes in Oakland and other cities that have gone forward with slow streets specifically, but heartened to see

达勒姆市的一位自行车和行人交通协调员 Dale McKeel 认为，投入更多的时间确实让他们设计出了一个能更好地满足社区需求的计划。他认为 Ortiz 对城市的贡献，以及她在推动社区参与方面的努力是值得称赞的。

对 Ortiz 而言，这个项目本质上是一个小范围内的尝试，帮助城市和那些持怀疑态度的市民建立对话窗口，从而建立信任。她认为，共享街道项目，规模有限、时效短暂。但她说："通过这个项目，人们会相信他们并不是只有一次机会表达意见，我们会一直保持倾听。"

疫情对行政预算造成了很大打击，政府想要做大这些项目，面临的主要挑战是保持对话窗口开放。与居民接触，进行对话并收集反馈意见，需要耗费大量的人力、时间和资金。因此政府与弱势社区之间建立信任可能还需要委托那些受信任的社区机构担任中介。Tamika Butler 和 Naomi Iwasaki 是交通方面的专家和社会公平倡导者，同时也是 NACTO 资助项目顾问。他们认为，达勒姆、亚特兰大和洛杉矶的案例告诉我们，即使知道问题会出现并扩大，政府也会倾听人们的诉求并确定实际的解决方案。

Butler 称，由社区组织发挥带头作用是充满挑战的。其中很重要的一环是我们需要倾听来自各方的意见，即使有些意见与交通问题并没有直接联系。但我们需要明白，在城市中，没有一个问题是孤立存在的。

城市规划的"创伤史"

"慢街"不仅仅是城市的一项建设政策——对城市规划者来说，它也是一个转折点。接受本文采访的几位社会公平倡导者认为，与他们合作的城市规划者和政府官员对相关问题已经形成了更深层次的理解。

Thomas 曾经在报纸上写过专栏文章评论奥克兰的"慢街"计划，她表示仍然对奥克兰和其他城市的最终实施结果感到失望，但也表示很高兴看到年轻一代的规划师正在学着与社区合作，这

that a younger generation of planners is learning to work with communities that go beyond the textbook. She said her op-ed has appeared on multiple university syllabi, and that she is working with professional planning groups to develop certification processes that better include people from different backgrounds.

12 To help build trust from the ground up, Thomas says, cities should invest in people who understand the communities they serve, and learn from the practices of social workers, counselors and mediators as they develop solutions.

13 "What is still missing is an interdisciplinary, multi-departmental approach: not just asking what we do with streets, but digging into how we make cities and communities healthier," she said. "If we shift to that focus, then our interventions will start to look a little different."

A slow streets barricade in Baltimore
(By David Dudley/Bloomberg CityLab)

14 "City planners think they just do bike lanes," he said. "But this is the industry that not that long ago rammed a bunch of freeways through neighborhoods and totally disconnected people. We need to reconcile that there is a history of trauma in what we do."

可是课本中从没教过的。Thomas 说，她的评论观点已被纳入多所大学的教学大纲，并且她正在与专业的规划团队合作，进行认证程序的研发，以更好地囊括不同背景的人们。

Thomas 认为，为了从根本上建立信任，城市规划者应该在那些了解其所在社区的社区工作者身上投资，并在制定解决方案时向社会工作者、社区咨询师和邻里调解员学习实践经验。

Thomas 还表示，我们现在缺少的是一种跨学科、跨部门的工作方法：不仅要问我们对街道做了什么，还要深入研究我们如何做才能让城市和社区发展更加健康。如果我们将重心转移到此，那我们的介入措施才会开始发挥作用。

巴尔的摩"慢街"路障牌
（摄影：David Dudley/Bloomberg CityLab）

"城市规划者认为他们只是在建设自行车道，"他说，"但就在不久之前，这个行业还建设了大量穿过社区的高速公路，将人们完全分离开来。我们不得不承认，我们对这个城市所做的正在叙述着它的"创伤史"。

15 As Oakland Slow Streets continues to evolve, his new objective is to help neighborhood groups take more leadership over micro-scale traffic interventions—for example, if one community wants a parkette on their block, while another wants a traffic circle, he wants the city to be able to supply safety tools and support for residents to make it happen on their own. It's a vision of DIY urbanism that reflects lessons learned from Slow Streets, as well as the austerity required by Oakland's $62 million budget hole.

奥克兰"慢街"政策在继续执行，新的目标是帮助社区组织提高领导力，完成小微尺度的交通干预——例如，一个社区想在自己的地块中修建一个小花园，而另一个社区想要一个交通环岛，Logan 希望城市能够支持居民自行操作并提供相应的安全工具。这是 DIY 城市主义的愿景，也是从"慢街"中得到的经验，同时也是奥克兰 6200 万美元预算缺口带来的财政紧缩的要求。

（文献来源：Laura Bliss. "Slow Streets" Disrupted City Planning. What Comes Next. https://www.bloomberg. com/news/articles/2021-01-06/the-swift-disruptive-rise-of-slow-streets. 杨慧、李蔓林、沈德瑶翻译）

第 5 课　城市韧性
Lesson 5　City Resilience

导读 ｜ Introduction

我们看待自然的方式 ｜ The Way We Treat Nature

　　我们看待自然的方式一直影响着我们的行动，尤其体现在我们与周围环境的关系中。由于经济机会的增加，以前难以获得的商品现在变得容易得到。我们生活在一个"开采—制造—使用—丢弃"的线性文化中，包括建筑业在内，这种现象非常普遍。在整个行业中，建设项目常见的做法是清除场地上的既有建筑和野生动物，或直接规划新的场地，而不是以场地为前提条件来调整项目。推倒重建已经成为常态，从广泛的意义上说，这种线性的建设方式损害了环境。

　　我们看到新建项目（城市空间扩张）遍布世界各地，由此带来了高昂的气候成本，这是不可持续的，同时也导致了单调的城市环境。建筑业的碳排放水平（建材生产占 9%）高于国际航空和海运的碳排放总和。

　　本单元描述了如何采取循环行动来减缓和应对气候变化，同时描述了这种行动的美学效果——循环设计如何帮助所有市民创造独特的城市环境。本单元阐述了循环行动的背景以及遵循自然原则的整体战略。循环行动的重点在于物质空间结构及其周围环境，以及物质环境与循环的相互影响。

预备知识 | Preliminary Knowledge

气候行动的关键概念与紧迫性 | Key Definitions of Climate Action and the Urgency

关键概念

Carbon 碳	Shorthand for all greenhouse gases, quantified in tonnes of carbon dioxide equivalent (tCO$_{2e}$) 所有温室气体的简称，以吨二氧化碳当量 (tCO$_{2e}$) 为单位进行量化
Downcycling 降级处理	When materials are recycled into another material of lower quality. 材料被回收利用为一种质量较低的材料
Recycling 回收利用	When materials are processed into products or materials used for their original or other purposes. 材料被加工成用于原产品或其他目的的产品或材料
Upcycling 升级利用	Recycling materials into new products or materials of better quality. 材料被回收加工成新产品或质量更好的材料
Aesthetics 美学	Visual and sensory qualities and the citizen's reflections to such qualities. 视觉和感官品质以及市民对这些品质的反馈

气候行动的紧迫性

虽然全球变暖的问题已经为人所知，但30多年过去了，它并没有任何重大改变。事实上，人类活动造成的全球碳排放一直在增加，导致大气中的碳浓度水平持续上升。只有在2020年，由于新冠肺炎疫情的影响，排放量才大幅下降（2020年上半年下降8.8%），但疫情的长期影响尚未明确。

由于全球变暖，我们已经目睹了平均温度和极端温度的上升、冰川的融化、海平面的上升和海洋的酸化、地下水位的下降，以及动植物栖息地的变化。气候变化使我们的生物多样性生态系统正遭受威胁，也对人类社会产生了重大影响。未来十年，城市人口将增长近 25 亿。未来十年关于城市和城市基础设施的决策，将决定各国是走上繁荣和可持续发展的道路还是走向衰退和灭亡。正如联合国指出的，我们需要在 2020 年至 2030 年期间每年减少 7.6% 的碳排放量，以使得全球变暖限制在 1.5℃ 以内。

为了实现《巴黎气候协定》的目标，建筑行业也必须寻求尽可能地减排，因为建筑业的碳排放量在总排放量中所占的比例很大。

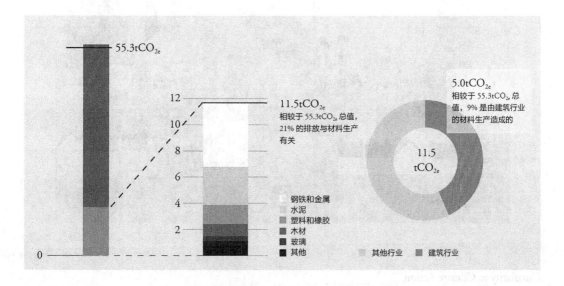

2018 年，全球碳排放总量为 553 亿吨二氧化碳当量。
21% 的排放与材料生产有关，9% 是由建筑行业的材料生产造成的。

课文讲解 | Text Explanation

Climate Action-going Circular Booklet

Circularity as Climate Action

1　Circularity means that resources should remain in an economic and functional cycle as long as possible. The concept of waste is not part of such a practice, as all materials are considered a resource.

2　Implementing[1] a circular approach is crucial to reducing carbon emissions[2] in the construction sector. In 2020, the EU[3] published a *Circular Economy Action Plan*. This plan provides a future-oriented agenda for achieving a cleaner and more competitive Europe in co-creation with economic actors, consumers, citizens and civil society organizations.

3　Circular measures provide us with solutions to mitigate[4] climate change by reducing carbon. Circular

走向循环——城市气候行动手册

气候行动的关键——循环

循环意味着资源应尽可能长时间地保持在一个经济和功能循环中。在这种理念下，已不再有"废物"的概念了，因为所有材料都被认为是一种资源。

实施循环方法对减少建筑行业的碳排放至关重要。2020 年，欧盟发布了《循环经济行动计划》。该计划提供了一个面向未来的议程，以期在与经济活动参与者、消费者、市民和民间社会组织的共同创造中，实现一个更清洁、更有竞争力的欧洲。

循环措施为我们提供了通过减少碳排放来缓解气候变化的解决方案。循环措施也使我们能够

measures also enable us to manage the consequences of climate change, such as making a city robust [5] and prepared for heavy rainfall. These measures provide solutions without necessarily reducing people's standard of living. In fact, it turns out that such solutions actually add value.

The Aesthetics of Circularity

4   The *European Landscape Convention* [6] highlights landscape planning as "strong forward-looking action to enhance, restore or create landscapes". This means that there is great potential to promote the desired landscapes at an early planning stage. But the right decisions must be taken.

5   Circularity offers a **narrative** [7] about how a cycle of life **unfolds** [8] in a broad sense, embracing everyone and everything. Every generation is a part of the cycle of life, a cycle formed by a continuity of human traditions and heritage. This narrative helps citizens to understand the **implications** [9] of climate change. And approaching nature with **humility** [10] might be a reason for implementing circular measures that can mitigate climate crises.

6   The influential landscape planner Ian McHarg articulated **"design with nature"** [11] as a version of circular planning. Circular strategy is an applied **mindset** [12] that follows the principles of nature. In this regard, circularity means adapting any human construction to the site-specific processes of nature. This aesthetic **framework** [13] provides **scope** [14] to prioritise beauty and the creation of places we **strive** [15] to take care of in the future.

Circular Urban Design

7   The circular design of our cities involves managing the physical environment within the site-specific processes of nature. But how does this relate to how we design our cities? And what's in it for the citizens?

8   From a circular **perspective** [16], it is important to see the urban structure as a whole—which requires more visionary, **holistic** [17] planning.

应对气候变化的后果，比如建设稳健的城市，做好应对暴雨的准备。这些措施提供的解决方案不一定会降低人们的生活水平。事实上，结果证明这些解决方案提高了生活品质。

循环的美学

《欧洲景观公约》强调景观规划是"提升、恢复或创造景观的强有力的前瞻性行动"。这意味着很可能通过早期规划推动理想景观的实现，但前提是必须决策正确。

循环经济从广义上解释了生命的循环，包含了所有人、所有事。每一代人都是生命循环的一部分，而这个循环就是指人类传统和遗产的延续。这种解释有助于市民理解气候变化的影响。而以谦卑的态度对待自然，可能是实施循环措施以缓解气候危机的一个原因。

伊恩·麦克哈格（Ian McHarg）是一位颇具影响力的景观规划师，他认为，循环规则就是要让"设计结合自然"。而循环战略则告诉我们从应用上如何遵循自然法则。在这方面，循环意味着人类的任何建筑都应适应特定场地的自然条件。这种美学标准为我们提供了一种视角，即优先考虑场所的美观度及创造性，这也是我们在未来需要重点关注的。

循环的城市设计

城市的循环设计是指在适应特定场地自然条件的前提下管理物理环境。但这与我们如何设计城市有什么关系？对市民来说又有什么好处？

从循环的角度来看，从城市的整体结构来进行规划是非常重要的——这种规划更有远见、更有历史意义。

9   But today's laws and regulations on the reuse of building materials are demanding. As a result, building materials end up in containers [18] even though they are fully usable. However, there is a call for a green change and a new rulebook in the construction industry among architects and contractors [19].

10 Circular actions are the results of choices made at a policy level, by individuals and at different stages in between. According to the EU Commission's *Circular Economy Action Plan*, the circular economy will provide citizens with high-quality, functional and safe products that are efficient and affordable, last longer and are designed for reuse, repair and high-quality recycling. A whole new range of sustainable services, product-as-service [20] models and digital solutions will bring about a better quality of life, innovative jobs and upgraded knowledge and skills. When we consider models of circularity, we have to adapt and understand them in the context of urban design.

11 Property [21] developers are responsible for the physical structures people live in. It is important to see the urban structure as a whole – which requires more visionary, holistic planning. The role of the developer seems to be shifting [22] towards being part of a long-term arrangement. And a circular approach to development is central.

The Built Environment

12 Buildings are an essential part of our living environment and provide space to work, socialize, sleep, or simply enjoy life. Building structures, walls, roofs and floors are resources that enable these operational [23] environments. However, as time goes by, cities and districts [24] evolve. Buildings need maintenance [25], people move in and out and grow older, and new generations present different needs.

13 Changing demands therefore require our buildings to transform from time to time. But producing new building products and materials causes carbon emissions, which again increases climate change. As a result of creating materials for use in construction, emissions are in many cases at least as large as for energy use throughout the building's life cycle.

但是现在的法律法规对建筑材料的再利用要求很高。因此，即使建筑材料是完全可用的，但最终还是被束之高阁。但现今已有呼吁，希望建筑师和承包商进行绿色改革，制定新的行业规则。

循环行动是个人在不同阶段基于政策作出各种选择的结果。根据欧盟委员会的《循环经济行动计划》，循环经济将为市民提供高质量、功能良好、安全的产品，这些产品高效、经济、耐用，并具有可循环使用、可修复、进行高质量回收的设计特点。可持续服务、"产品即服务"模式和数字解决方案这一系列全新的措施将提高生活质量，带来富有创意的工作机会和更先进的知识与技能。当考虑循环模式时，我们必须根据城市设计来调整和把握。

房地产开发商对人们居住的物理环境负有重要责任。从整体上规划城市结构很重要——这种规划更有远见、更有历史意义。开发商的角色似乎正在发生转变，成为助力城市长远发展的参与者，而长远发展的核心则是可循环发展。

建成环境

建筑是我们生活环境中必不可少的一部分，为我们提供了工作、社交、休憩、享受生活的空间。建筑的结构、墙体、屋顶和楼板则是打造这些使用场所的资源。然而，随着时间的推移，城市和街区也在发生相应改变：建筑需要修缮，人们搬进迁出，越来越年长，而新一代的居民又提出了不同的需求。

需求的发展变化也要求我们经常对建筑进行改造。但是生产新的建筑产品和材料将导致碳排放增加，加剧气候变化。建筑材料的生产所带来的排放量，在许多情况下至少与建筑整个生命周期中的能源使用量相同。

14　The EU stipulates [26] that starting in 2020, 70% of all construction waste must be recycled. It is crucial that we constrain all actions that escalate [27] climate change. We must realize that each brick [28], wall, door and windowpane [29] has a value.

Reusing Existing Buildings

15　Preserving and extending the lifetime of existing buildings normally results in less carbon emissions than demolishing [30] and building new ones. To reduce emissions created during the production of building materials, existing buildings should be reused where possible and space efficiency maximized to avoid building more.

16　In the Økern area development in Oslo, an 18-storey building from 1970 is planned to be reused in the future. Calculations made with the software One Click LCA show that if the building is demolished, and a new one rebuilt, the materials will account for approximately 4,000 tonnes [31] of carbon. By contrast, preservation and rehabilitation [32], which retain most of the heavy structures [33], will emit about 1,900 tonnes of carbon. This contributes to a reduction in emissions of approximately 55%.

17　Another example of the same approach is found in Finland. The City of Helsinki aims to achieve carbon neutrality [34] by 2035. This target is ambitious and requires implementation of a wide range of actions. In an area where most of the buildings were constructed during the 1960s and 1970s and are now in need of heavy refurbishment [35], Sweco [36] has performed construction calculations indicating that it is possible to achieve even up to 59% lower embodied carbon [37] while providing preserving refurbishment. Maintaining the existing bearing constructions in steel and concrete in particular had a huge impact on lowering the carbon footprint. Moreover, the construction cost analysis indicated that a preserving renovation would be more affordable than the construction of new buildings.

Circular Actions for Constructions in Existing and New Areas

18　In general, for existing areas it is better to prolong the lifecycle of and reuse complete buildings or elements rather than recycle mass from demolished buildings. Recycling is better than downcycling, specifically in

　　欧盟规定，自 2020 年起，70% 的建筑垃圾必须回收利用。对所有会加剧气候变化的行为加以限制，这一点至关重要。我们必须意识到每一块砖、每一面墙、每一扇门、每一个窗格都是有回收价值的。

既有建筑的再利用

　　维持和延长既有建筑的使用寿命，相较于拆除后新建，通常可以减少碳排放。为了减少建筑材料生产过程中的碳排放，我们应该尽可能地重新利用既有建筑，并使空间效率最大化，以避免新建更多的建筑。

　　在挪威奥斯陆市的 Økern 区，一幢建于 1970 年的 18 层建筑将被重新利用。根据 One Click LCA（自动化生命周期评估）软件计算，如果拆除该楼，原地重建新楼，将会产生大约 4000 吨的碳排放。相反，保留其大部分的重型结构并进行修复，只会产生大约 1900 吨的碳排放，减少了大约 55% 的碳排放。

　　在芬兰也有案例使用了相同的方法。赫尔辛基市致力于在 2035 年前实现碳中和的目标。这个目标雄心勃勃，需要实施一系列的行动来实现。在该市某区域，大部分建筑建于 20 世纪六七十年代，如今需要大刀阔斧地翻新改造。Sweco 公司通过施工计算发现，如果保留这些建筑并翻修，可能减少高达 59% 的隐含碳排放。尤其是保留现有的钢筋混凝土承重结构，将对减少碳足迹产生极大的影响。而且，施工成本分析表明，以保留和再利用为主的翻新改造比拆除新建更加经济适用。

建成区域和新建区域的建设循环行动

　　通常，对建成区域来说，相较于从拆除的建筑中回收材料，更好的做法是延长既有建筑或其构件的生命周期并重复利用。回收利用优于降级处理，特别是考虑到改变过程中所需能源的情况。

regard to the energy needed in the transformation process. When planning development of new areas one should think of what kind of design concept is best in each case. Developing new areas provides the opportunity to design with the end of the lifecycle in mind.

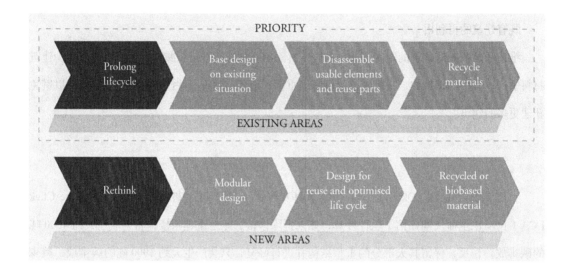

Reusing Building Elements

19 When the entire structure cannot be adequately reused, reusing existing building elements makes a good **alternative** [38]. A positive example of the potential extensive reuse of building materials and elements is Construction City, an alliance of construction businesses in Ulven, Oslo.

20 Global cement production is responsible for approximately 5% of total carbon emissions. This is more than the global emissions from air travel. So there is no question that there is huge potential to reduce the carbon footprint of the project when it comes to concrete. Therefore, the main focus has been on concrete. The project has an opportunity to reuse, or recycle, 70–90%. They have the potential to reuse 3,400 tonnes of concrete by reusing and **reconfiguring** [39] elements, and as a result reduce overall carbon emissions by 952 tonnes. This is the equivalent of about 350 **round-trip** [40] flights from London to New York. The client further requires a minimum 100-year **lifespan** [41] for the structure. And to maximise end-of-life reuse, the concrete contractor has to create a plan showing how they will disassemble the elements.

对新建区域来说，制定发展规划时应考虑每个项目最适宜的设计理念，为我们提供了在设计中考虑全生命周期的机会。

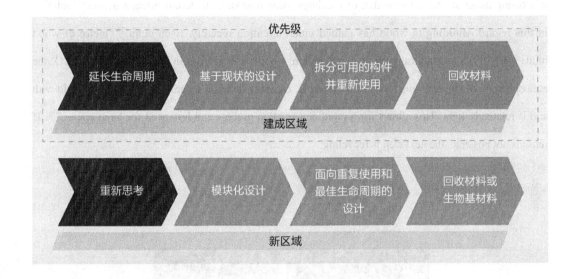

重新使用建筑构件

当整个建筑结构无法完全重复使用时，再利用其构件是一个不错的替代方案。奥斯陆 Ulven 的建筑行业联盟"建设城市（Construction City）"，就提供了广泛再利用建筑材料和构件的积极案例。

全球水泥生产的碳排放量约占碳排放总量的 5%，这比全球航空运输产生的碳排放量还多。由此看来，从混凝土入手来减少该建筑项目的碳足迹，无疑有着巨大的潜力。因此，该项目的重心就放在了混凝土上，70%~90% 的建筑构件能再次使用或回收利用，通过再利用和重置可以实现 3400 吨混凝土的再利用，最终减少 952 吨的碳排放量，这相当于约 350 个从伦敦到纽约的往返航班的碳排放量。委托方后来提出了要求，建筑结构至少要有 100 年的使用寿命，为了最大化全生命周期结束后的重复使用率，混凝土承包商还不得不制作了一个方案，详细说明如何拆解混凝土构件。

Recycled and Biobased Materials

21 To reach carbon neutrality targets, we need to find ways to produce buildings with minimal carbon emissions and to maximise operational energy efficiency during the use phase. In addition to the reuse of existing materials, the construction of buildings using renewable materials offers a **decent**[42] solution in this regard. Building in wood has gained much attention recently. Each **cubic**[43] metre of wood binds a ton of the **greenhouse gas**[44] CO_2. The more wood we store (e.g. in buildings), the less CO_2 the atmosphere needs to deal with. Wood is renewable, can provide large separate components to build with and is relatively light. This allows for a rapid construction process and less need for large **cranes**[45]. If reuse is not possible, the next step is recycling.

Generally, up to 50% carbon is saved by building in wood compared to conventional construction.

回收材料和生物基材料

为了达到碳中和的目标，我们在建造过程中要设法将碳排放量降至最低，同时最大限度地提高使用阶段的能源效率。除了对现有材料的再利用外，施工中使用可再生材料也是一个良好的解决方案。近些年来，木结构建筑备受关注。每立方米的木材吸收一吨温室气体（二氧化碳）。我们储存的木料越多（比如在建筑中），大气中需要处理的二氧化碳就越少。木材是可再生的，可以为建筑提供大型的独立构件，而且重量较轻。这种材料特性也能满足快速施工的要求，减少对大型塔吊的需求。如果不能重复使用，还可以回收利用。

一般来说，与传统建筑相比，以木头为原材料的建筑能节省多达 50% 的碳排放量。

Designing for Reuse

22 The reuse of existing buildings is another method that prefers reuse over new builds. Entire buildings can be moved from one site to another.

23 Norway has a long tradition of moving its houses. In fact, the dismantling and rebuilding of log cabins in new locations is part of Norwegian building history. In the future, when designing new buildings or structures of various kinds, the same principles should be taken into consideration.

24 Today, solutions already exist to make movable buildings. Parts of buildings can be detached without destroying the main structures. Elements of lightweight wooden structures are quick to assemble on site and can then be moved somewhere else or even implemented as a part of another building. A good example of this **modular design** [46] is by a Finnish building **supplier** [47] that provides **prefabricated** [48] learning spaces that can be dissembled and moved to another site when needed. Schools, preschools, gyms and cantinas are made like this. With this approach, it is possible to maintain entire or parts of buildings and **relocate** [49] them somewhere else. There is no need to demolish in the usual sense of the word.

为重新使用而设计

与新建建筑相比，对既有建筑的再利用是一种更受欢迎的设计方法。整栋建筑能从一处转移到另一处。

挪威有着悠久的"建筑搬家"传统。事实上，拆除小木屋并在新场地进行重建本身就是挪威建筑历史的一部分。未来，设计新建筑或各式结构时，也应考虑再次利用原则。

如今，可移动建筑已有办法实现。在不破坏主体结构的情况下，我们可以拆下建筑物的部分构件。轻质木结构的组件可以在现场快速完成组装，然后转移到其他场地，甚至可以用于另一处建筑的构建。这种模块化设计的一个典型案例就是芬兰建材商提供的预制学习空间，它可以随时拆卸并进行转移。学校、幼儿园、健身房和小酒吧都可以按照这种方式建造。有了这种方法，就可以保留整个建筑或其中一部分，并将其搬迁到其他地方。如此一来，也就没有必要进行通常意义上的拆除了。

（文献来源：SWECO. Urban Insight Report Going Circular:A Vision for Urban Transition.https://www.swecourbaninsight.com/wp-content/uploads/2020/12/Summary-of-report-Going-circular.pdf. 余铭航、吴梦洋、陆建、张浩然、林若然翻译）

词汇 | Vocabulary

[1] implement *vt.* 使生效，贯彻，执行；*n.* 工具，器具，用具

[2] emission *n.* 排放物，散发物（尤指气体）

　　carbon emissions 碳排放

[3] European Union（EU） 欧洲联盟（欧盟）

[4] mitigate *vt.* 使缓和，使减轻；*vi.* 减轻，缓和下来

[5] robust 强健的，健康的，粗野的，粗鲁的

[6] *The European Landscape Convention* 《欧洲景观公约》

认为景观没有必要划分为"自然的"或"文化的"两类，因为欧洲所有的景观都已经或多或少地受到人类的影响。因而"文化景观"的术语显得过于累赘，所有欧洲景观都具有不同程度的文化性。

[7] narrative *n.* 记叙文，故事、叙述，叙述部分；*adj.* 叙述的，叙事体的

[8] unfold *vt.* 打开，呈现；*vi.* 展开，显露

[9] implication 暗示；牵连；卷入；含义

[10] humility 谦卑，谦逊

[11] design with nature 设计结合自然

[12] mindset 倾向，心态

[13] framework 结构，框架

[14] scope 范围；空间；眼界

[15] strive 努力

[16] perspective 看法；角度；观点

[17] holistic 全部的

[18] container 器皿，容器

[19] contractor （建筑、监造中的）承包人

[20] product-as-service *产品即服务*

[21] property 此处意为所有权、房地产

[22] shift 转变

[23] operational 可使用的

[24] district 区域；行政区

[25] maintenance 维修

[26] stipulate 约定；规定

[27] escalate 逐步升级，逐步扩大

[28] brick 砖块

[29] windowpane 窗玻璃，这里指窗格

[30] demolish 摧毁，推翻，拆毁（尤指建筑物）

[31] tonnes [计量] 吨（tonne 的复数）

[32] rehabilitation 修复，翻新

[33] heavy structures 重型结构

[34] carbon neutrality 碳中和

[35] refurbishment 翻新

[36] Sweco 知名工程和建筑咨询公司，关注未来可持续发展的社区和城市

[37] embodied carbon 隐含碳排放

[38] alternative 可替代的事物，文中意为"选择"

[39] reconfigure 重新装配，改装

[40] round-trip 往返行程

[41] lifespan 预期使用期限；生命周期

[42] decent 得体的；良好的

[43] cubic 立方的；立方体的

[44] greenhouse gas 温室气体

[45] crane 鹤；在建筑行业中指起重机，吊车

[46] modular design 模块化设计

[47] supplier 供应商

[48] prefabricated （建筑物）预制构件的

[49] relocate 迁移

练习与思考 | Comprehension Exercise

1. 第 3 段中阐述了 "These measures provide solutions without necessarily reducing people's standard of living. In fact, it turns out that such solutions actually add value. " 你认为此处 "value" 包含哪些方面?

2. 第 10 段中提到 "A whole new range of sustainable services, **product-as-service** [20] models and digital solutions will bring about a better quality of life, innovative jobs and upgraded knowledge and skills. When we consider models of circularity, we have to adapt and understand them in the context of urban design." 请解释 "product-as-service" 的含义,并在城市设计领域中阐释你的理解。

课后延伸 | Reading Material

城市热岛 | Urban Heat

Urban Heat

1 Recent years have brought new high temperature records across Europe, with the summer of 2019 witnessing record highs across France, the Netherlands, the UK, Belgium and Germany. These hot days might be enjoyable for some, but they also pose a significant and growing health risk to our families and communities.

2 Certain groups in cities are more vulnerable to heatwaves than others, such as young children and older people. While the proportion of younger people in Europe is not set to change, Europe is getting older, and the elderly are particularly at risk. In Europe, the number of people over 65 is rapidly increasing, with those over 80 expected to more than double from 5.6% in 2018 to 12% by 2060.

3 Cities will face multiple shocks and stresses in the future, with the vulnerable usually paying the highest price. The coronavirus pandemic is also bringing this to light. We need to manage how we develop our buildings and public spaces in our cities to protect our most vulnerable.

Impacts on Health

4 Extreme heat events are linked to a multitude of health impacts : dehydration, heatstroke and exhaustion, increases in incidence of disease (cardiovascular, respiratory and cerebrovascular) and increases in premature death. The elderly, infants and young children, those with pre-existing health problems and those in hospital or bedridden are most at risk. While these groups are physiologically more sensitive to heat, how people subjectively experience heat is also important and depends on a complex interaction of many factors:

5 Physiological. Previous long periods of hot weather can result in physical acclimatisation, reducing the negative impact on the human body.

Local climate. Humidity, air temperature, shading and windchill all impact on how heat is experienced.

Socio-economic. Being socio-economically disadvantaged affects your experience of extreme heat—for example, due to the quality of available housing, the availability of affordable mitigation measures like air conditioning, and proximity to green spaces.

城市热岛

近年来，欧洲各地出现了新的高温纪录，2019 年夏天，法国、荷兰、英国、比利时和德国各地都出现了创纪录的高温。这种高温天气对某些人来说可能是愉快的，但同时也对我们的家庭和社区构成了重大的、日益增长的健康风险。

城市中的某些群体更容易受到热浪的影响，如幼儿和老年人。虽然欧洲的幼儿比例不会改变，但欧洲正面临老龄化，老年人尤其面临风险。65 岁以上的人口正在迅速增加，80 岁以上的人口预计将增加不止一倍，从 2018 年的 5.6% 增加到 2060 年的 12%。

城市在未来将面临多重冲击和压力，弱势群体通常首当其冲。新冠肺炎病毒的大流行也让人们看到了这一点。我们需要思考如何发展城市的建筑和公共空间，以保护最脆弱的群体。

高温对健康的影响

极端高温天气会带来众多健康问题：脱水、中暑和衰竭、疾病发病率增加（如心血管疾病、呼吸道疾病和脑血管疾病）以及早逝人数增加等。这些健康问题对老年人、婴幼儿、既有疾病人群、住院或卧床病人来说，风险是最高的。虽然这些人群在生理上对高温更敏感，但他们对高温的主观感受也非常重要，受到各种因素的互相影响。

生理因素。长期生活在炎热天气中可造成生理适应，会减轻高温对人体的负面影响。

局部气候。湿度、空气温度、遮挡物和低风量都会影响人体对高温的感受。

社会经济因素。社会经济条件差会影响人们对极端高温的感受——例如住房条件不好、负担不起空调等降温设备、附近没有可去的绿地等。

Psychological. Spending time in urban greenery can improve perceived well-being, alleviating the perception of thermal discomfort.

Urban Heat Island Effect

6 The impact of heat in cities is exacerbated by the urban heat island effect (UHI). This is a phenomenon whereby many cities are warmer than their rural surroundings due to a combination of dark surfaces, high-rise buildings, lack of green space, lack of wind and air pollution trapping heat from the sun, industry, traffic, and day-to-day activities. This effect is most pronounced at night, with extreme cases displaying differences of 5–10 °C between rural and urban areas, with the average effect a difference of 2–4 °C. Heatwaves also make the UHI effect worse, and this effect will intensify with climate change.

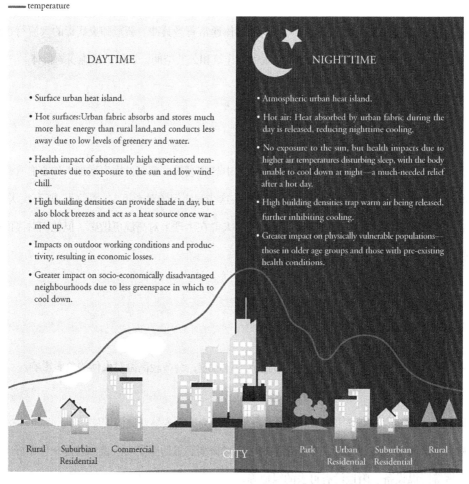

Daytime and nighttime impacts of the UHI effect, as well as the health risks due to heat.
These result in an uncomfortable urban environment – one with reduced liveability and spatial quality.

心理因素。待在城市绿地可以提升幸福感，缓解热不适感。

城市热岛效应

城市热岛效应（UHI）加剧了城市极热天气的影响。城市热岛效应是指许多城市的气温高于周边农村地区的现象，其原因是城市深色表面面积大、高层建筑林立、绿色空间缺乏、空气流动性差以及空气污染严重，这些因素都会阻止因太阳、工业、交通和日常活动而产生的热量的释放。这种效应在夜间表现最为明显，农村和城市地区最多会产生 5~10 ℃ 的温差，而一般情况下的平均温差为 2~4 ℃。热浪也会加剧城市热岛效应，而这种效应会随着气候变化而加剧。

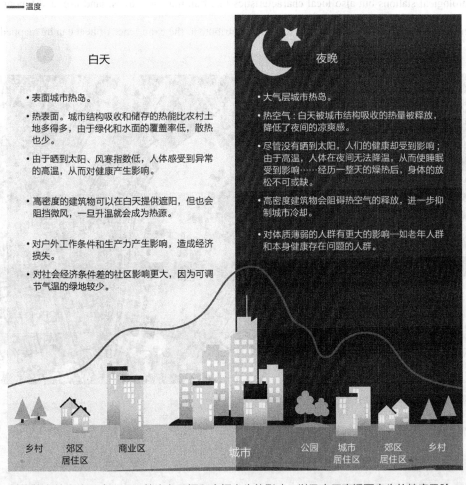

城市热岛效应的影响：UHI 效应在日间和夜间产生的影响，以及由于高温而产生的健康风险，导致了城市环境愈发不舒适——降低了宜居性和空间质量。

Approaches to Building Resilience

7   The first approach to building resilience is: be prepared. Data is crucial for increasing the capacity to prepare for challenges. In a city with a strong capacity to prepare, the municipality has conducted research on important thresholds and criteria particularly for vulnerable groups. Risk and vulnerability assessments pinpoint the particular threat to our 2-year-old and 80-year-old.

Heat Map of Utrecht, the Netherlands

8   The Netherlands has a standardised method for heat maps, making them comparable among different municipalities. This method is based on calculating the Physical Equivalent Temperature (PET), which includes weather parameters like wind, air temperature, humidity and radiation from nearby meteorological stations but also local characteristics like buildings and trees, land use, and topography. This means that many of the different factors that contribute to the experience of heat can be mapped.

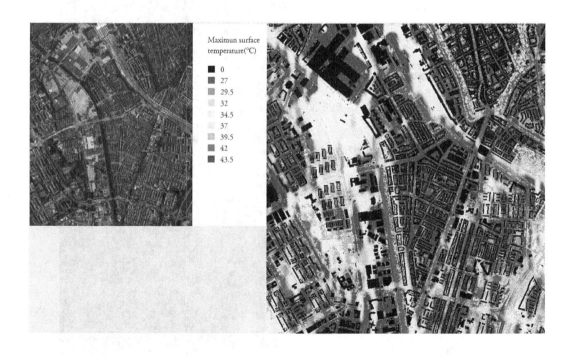

建立韧性的方法

韧性建设的第一个方法是：做好准备。数据对于提高应对挑战的能力是至关重要的。在应对力强的城市，政府已对重要的阈值和标准进行研究，尤其是针对弱势群体的阈值和标准。风险和脆弱性评估显示 2 岁婴幼儿和 80 高龄老人面临特殊威胁。

荷兰乌得勒支热力图

荷兰有一套标准化的热力图制作方法，可以在不同的城市之间进行比较。该方法基于物理等效温度（PET）的计算，包括风、空气温度、湿度和附近气象站的辐射等天气参数，以及建筑物和树木、土地利用和地形等当地特征。这意味着，造成热感受的许多不同因素都可以被绘制出来。

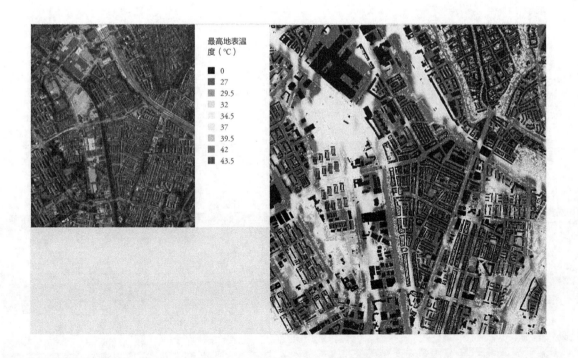

最高地表温度（℃）
- 0
- 27
- 29.5
- 32
- 34.5
- 37
- 39.5
- 42
- 43.5

（文献来源：SWECO. Urban Insight Report Building Resilience: Being Young and Getting Old in a Hotter Europe. https://www.swecourbaninsight.com/wp-content/uploads/2020/10/report-building-resilience.pdf. 李玘苏、王琛芳、陆建、杨潇晗、王畅、章思予、杨蕊源翻译）

第 6 课　城市社区与公共空间

Lesson 6　Urban Community and Public Space

导读 | Introduction

城市健康与人民福祉 | Urban Health and Well-being

据联合国统计，预计到 2050 年，全球约 70% 的人口将居住在城市。城市密度将继续增加，城市规模将持续扩大。如果我们设计城市的方式没有随着居民数量的增加而改变，未来将只有少数人愿意继续居住在城市。

我们需要重新思考我们的城市规划。

尽管城市正在倡导自行车和步行的出行方式，但城市规模的扩大、区域连通性的增强正刺激着长途出行的需求，使得机动车成为最便捷的选择。因此，实现可持续发展，不仅需要改变出行方式，也需要改变城市规划。

要实现交通的可持续发展、打造健康宜居的城市、保障人民福祉，应从交通规则、城市规划、社会学等不同角度进行综合考虑，将工程科学与社会科学有机结合。

城市街道设计和技术升级调整不仅有利于我们减少碳排放、减少交通事故、减轻交通压力，也有利于城市规划者和建筑师以新的方式思考城市的发展，实现人与服务的和谐统一。城市建设者应重视邻近性，重新设计街道空间，以实现城市居民社交需求为目的，让所有市民对城市空间充满信心。

预备知识 | Preliminary Knowledge

一体化社区规划方法 | Symbiocity Community Planning Approach

为人口密集的社区提供高品质、有韧性的生活环境，需要一套专门的一体化方法。

瑞典的"共生城市"一体化社区规划方法（Symbiocity Community Planning Approach）为跨学科、能实现可持续发展的城市规划制定了框架和标准，重点关注公众参与性和能力建设，强调规划过程中利益相关者的共同创造，这种方法已在许多国家实施应用。

再如德国的欧洲一体化城市发展方案，同样强调公众参与性，并与国家资金挂钩。

以市民为中心的瑞典"共生城市"一体化社区规划方法

（图片来源：SWECO, Urban Insight Report_Neighbourhoods of tomorrow，杨慧翻译）

　　城市和建筑的设计图纸、视觉效果、潜在解决方案及成功案例，都有助于启发和帮助社区定义其愿景、鼓励合作、共同创造、确定权属。在既有城市规划和宜居性建设领域，已经有许多成功的案例可供参考。

　　根据一体化社区规划方法，以市民为中心是至关重要的。一体化社区规划方法与其他整体性规划方法是通过利益相关者的积极参与来设定愿景、指导规划、实现预期结果的。

　　同样，解决城市致密化（densification）、气候韧性和宜居性等相关问题，也需要采用可视化方法以及有效的深度合作。

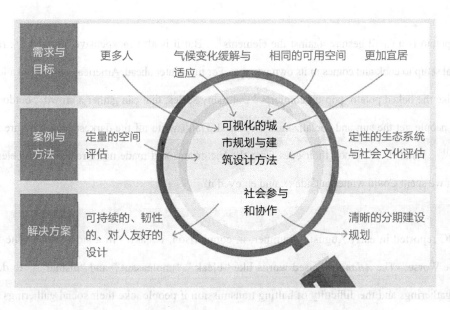

应对挑战——解决城市致密化、气候韧性和宜居性相关问题，
必须采用可视化方法以及深度并有效的合作。
（图片来源：SWECO, Urban Insight Report_Neighbourhoods of tomorrow, 杨慧翻译）

课文讲解 | Text Explanation

How to Make the Most[1] of Covid Winter

1   In Victorian England, the baked potato had dual purposes. Sold from **street-side**[2] "cans"—metal boxes on four legs, with charcoal-fueled fire pots within—the potatoes could be used as hand-warmers when tucked inside a mitten or muff, or body warmers when consumed on the spot as a hot and filling snack. Potato sellers by the hundreds set up cans on London streets, selling their wares from August to April, as **ubiquitous**[3] as today's ice cream trucks but serving the opposite season.

2   Buyers and sellers of **those spuds had little choice**[4] but to be out on the streets, whatever the weather. That's where the business was, that's where workers could buy a quick meal, that's where friends might encounter each other, standing close to the can for warmth. A little food, a little fire, a little chat—these elements made being outdoors in winter **bearable**[5].

3   A hot potato is a small **gesture against the elements**[6]. But it is also inexpensive, **portable**[7], requires minimal setup to cook and comes in its own wrapper. For the winter ahead, American cities need a lot more ideas like the baked potato: **pop-up comforts**[8], at many scales, that can gather a **crowd**[9] outdoors and ensure people get the sun and socialization they need. **Don't write off the darkest season before it even begins**[10]. What if cities took their cues from the Victorians, and made no retreat from the elements? What if we spent Covid winter outside … and enjoyed it?

4   As CBC reported in early August, "Summer, in comparison, is easy." In a story titled "The Winter Will Be Worse," The *Atlantic*[11] used words like "bleak," "unpleasant," and "dismal"[12] to describe chilly gatherings and the difficulty of halting transmission if people take their social gatherings inside. But illuminated sunset walks, invigorating bike rides, and hot chocolate by the fire pit don't sound so terrible.

应对压力、保持活力：以城市公共空间的方式

维多利亚时代的英国，烤土豆有着双重作用。烤土豆在街边的"罐头筒"中售卖——一种四条腿的金属容器，内有木炭火盆——当你把烤土豆塞进手套或暖手筒中，它可以当作暖手炉；如果你当场吃掉，它又是一种热乎乎的零食，能填饱你的肚子，也能温暖你的身体。从 8 月到来年 4 月，数以百计的土豆小贩在伦敦的街道上摆好罐头筒，售卖他们的产品，就像今天的冰激凌车一样无处不在，只是烤土豆是在相反的季节供应罢了。

无论天气怎样，买土豆的人和卖土豆的人都能在街头交易。那是做生意的地方，是工人可以买快餐的地方，也是朋友们可以偶遇的地方。而且，靠近罐头筒，你就可以感受到温暖。一点儿食物，一点儿火苗，一点儿闲聊——就能使冬天的户外不那么难熬。

一个热乎乎的烤土豆虽微不足道，却作用不小。而且它价格便宜，便于携带，做起来简单，而且有自己的包装。冬天将至，美国的城市需要更多像烤土豆这样的创意：各种尺寸的移动售卖车（pop-up comforts），可以让人们聚集在户外，并获得所需要的阳光和社交活动。不要冬天还没有开始就对它不抱希望、放弃它。如果我们的城市可以从维多利亚时代获得启示，直面冬天恶劣的环境，不退却，那将会是何种景象呢？如果我们可以正视疫情并在户外活动……甚至享受这样的冬天，又将会如何呢？

正如 CBC 在 8 月初的报道中所说，"相比之下，夏天就好过多了"。美国《大西洋月刊》在一篇名为《冬天会更糟》的文章中，使用了"凄凉"（bleak）"不愉快"（unpleasant）和"沉闷"（dismal）等词来描述寒冬天气的聚会，同时也指出，人们在室内举行社交聚会将会增加疾病传播的风险，但想想夕阳下的散步、充满活力的骑行、火炉边的热可可，这些似乎还不错。

5　With wildfires in the Western U.S. and heat waves across the South, it's a **stretch**[13] for Americans to start thinking about winter. **But now is the time to think ahead**[14]. Many of the activities that have been keeping people in cities safe and sane since March—walks, bike rides, stoop gatherings, and park picnics—will literally take on a different **hue**[15] as temperatures drop, the wind picks up and the sun sets at 5 p.m. Urban winter in the many parts of the U.S. can take cues from counties with snowy slopes—but also from other cultures used to dealing with cold.

6　Scientists remain divided on why winter is flu season in the Northern Hemisphere, but two of the most popular hypotheses are increased time spent in sealed indoor environments and a lack of sunlight, which compromises the **immune system**[16]. Covid may work the same way, meaning that in 2020 we have two viral reasons to stay outdoors **rather than one**[17].

7　"The story I always tell about our winter city strategy, was that we, not just in Edmonton but across North America, felt we could respond to winter by making it go away," says Ben Henderson, councillor for Ward 8 on the Edmonton City Council and one of the authors of the city's seven-year-old **WinterCity initiative**[18]. "You got up in the morning and it was probably dark, you connected to your vehicle without going outside, you went to the **pedway**[19] and ate lunch without going outdoors. We were never getting direct sunlight or fresh air and then we wondered why we became depressed."

8　Edmonton is among Canada's coldest cities—though it is a dry cold city—with January temperatures reaching as low as 3 degrees Fahrenheit. So the idea was: If you can enjoy January outdoors there, you can enjoy it anywhere. By making it easier to get to work, play outside and find things to do outside the home, the WinterCity Strategy targeted both physical and mental health. By 2017, 44% of respondents to a WinterCity survey said their perception of winter in Edmonton had become more positive. U.S. cities have months, rather than years, to do the same for Covid winter, but the strategies shouldn't be that different. It has become clear that the safest place to be with other people is outdoors, and we can learn from winter cities how to keep the good times (literally) rolling.

美国西部野火肆虐，热浪席卷了整个南部，对美国人来说现在就开始考虑冬天的问题还有些勉强。但是时候要提前考虑了。自 3 月份以来的许多安全又理性的城市户外活动——散步、骑行、门口聚会及公园野餐——将随着气温下降、风力增强以及下午 5 点就下山的太阳而发生变化。在美国的许多城市，冬天的过法可以借鉴满是雪坡的县乡，也可以借鉴其他国家——有应对寒冬经验的国家。

对于为什么冬天是北半球的流感多发季节，科学家们仍然存在争议，但最可能的两个原因是：人们在封闭的室内环境中待的时间过长以及缺少阳光。这二者都会损害免疫系统。疫情也可能是这种情况，这就意味着 2020 年我们至少有两个理由要待在室外。

Ben Henderson 说："我经常讲到的，我们城市的冬季策略是：让冬季消失。这就是我们应对冬季的办法，不仅是埃德蒙顿，而是整个北美都该如此。"他是埃德蒙顿市八区的议员，是该市具有 7 年历史的"冬城倡议（WinterCity initiative）"起草者之一。"你早上起床的时候天可能还没亮，你无须走到室外就可以坐进小汽车中，你也不必非走到户外去吃午饭。我们总不去直接接触阳光或新鲜空气，却在纳闷我们为何会变得沮丧不已。"

埃德蒙顿是加拿大最冷的城市之一——尽管是寒冷干燥的天气——一月份的气温可以低至 3 华氏度（零下 16 摄氏度）。所以我们认为：如果你能忍受埃德蒙顿一月份的户外生活，你就能享受任何地方的冬天。这个城市提出了"冬城倡议"策略以促进市民身体与心理健康为目标，为市民外出工作、户外游玩以及组织户外活动提供便利和支持。在"冬城倡议"的一项调查中，截止到 2017 年，有 44% 的受访者表示他们感觉到埃德蒙顿变得更加积极有活力了。对于疫情下的冬季，美国的城市需要应对好几个月（但不是几年）。这些城市做法相同，但整体策略不该有那么大差异。有一点是非常明确的，那就是，与其他人共处的最安全的场所是户外。我们可以向冬城学习如何在冬季延续美好时光。

9　If Summer 2020 made best-sellers out of such backyard pleasures as potting soil, inflatable pools and sidewalk chalk[20], Winter 2020 should see a run[21] on long underwear, fire pits and faux fur cushions. Edmonton's WinterCity website includes actionable toolkits at multiple scales, starting with winter fashion. Dress in layers, invest in silk and wool long underwear, get over your prejudice[22] against parkas. Many people do this as a matter of course when gearing up[23] for a day of skiing or a turn around the ice rink. But in cities, people dress for the destination, not the journey. "People dress saying, I'm going from my home to this business. What's the least amount of clothing I can wear for the tolerance of walking x feet?" says Simon O'Byrne, senior vice president of community development for global design consultancy Stantec. "We have to switch that, and dress to loiter[24]."

10　O'Byrne, who is also co-chair of the WinterCity Advisory Council, adds, "**Stickiness encourages people outside**[25]. Moscow does year-round farmers markets. The artists' community has been **pulverized**[26] by Covid. As much as we can, we should embrace things to help the local artists' community." He suggests commissioning visual artists to illuminate dark spaces, via **murals or light installations**[27], and hiring musicians for distanced outdoor concerts.

11　Cities should also invest in places to loiter. All of those outdoor restaurants that are supporting local businesses and bringing liveliness back to the streets? In New York City, at least, they are scheduled to shut down at the end of October, while the **mayor and governor**[28] bicker over indoor dining. But citiesneed to catch up to ski areas, which long ago **figured out**[29] how to make après ski activities like outdoor bars and music venues as much of an attraction as the slopes. Wind breaks (with openings above and below for ventilation), patio heaters and sun orientation can all take outdoor dining further into 2020.

12　WinterCity's *Four Season Patio Design Tips* also include higher insulation value materials, like wood or straw bales rather than metal seating, as well as simple solutions like blankets, which offer customers the winter **equivalent**[30] of being able to reposition your chair in the sun—though that works year-round.

如果 2020 年的夏天能使后院中的乐趣变得流行，比如盆栽、充气泳池、人行道彩绘，那么 2020 年的冬天应该可以看到秋裤、火盆和人造毛坐垫的畅销。埃德蒙顿"冬城倡议"的网站中列出了各种规模的可操作工具包，即将引领冬季时尚。穿着讲究层次搭配，可随意增减，购买丝毛材质的秋衣秋裤，摆脱偏见穿上派克服等，这些都是人们打算去滑雪或冰上乐园时理所当然会做的着装准备。但在城市里，人们只会根据目的地的温度选择服装，而不考虑行程过程。"人们穿衣服的时候考虑的是，我要从家里出发到公司，需要走几英尺路，怎样能穿得最少？"全球设计咨询公司 Stantec 的社区发展高级副总裁 Simon O'Byrne 说，"我们得换个思路，为随时可能在户外漫步而准备衣物"。

O'Byrne 也是冬城倡议咨询委员会的联合主席，他补充道，"黏附力会鼓励人们到来到户外。这就是为什么莫斯科农贸市场全年都有，而艺术家团体却被新冠摧毁了。我们应该倾尽所能去帮助当地的艺术家团体。"他建议委托视觉艺术家安装壁画或灯光来点亮黑暗的空间，聘请音乐家隔空举办户外音乐会。

城市也应该投资适于休闲漫步的场所，比如那些既能支持本地经济又能恢复街道活力的户外餐厅？至少纽约市已计划在 10 月底关闭它们，尽管市长和州长对室内用餐问题还存在争议。城市需要向滑雪场学习，这些滑雪场早就知道如何设计滑雪之后的活动，比如户外酒吧和音乐场所等，让它们像滑雪坡道一样具有吸引力。防风屏障（上下两端都有通风口）、户外取暖器、朝阳的位置取向，这些都能使 2020 年的冬季户外用餐正常进行。

"冬城倡议"的《四季露天设计技巧》还提到，可以使用绝缘值较高的材料，如木材、稻草包来替代金属座椅，也可以使用像毛毯这样的简单物品，使消费者在冬天也能感受到置身阳光下的温暖——虽然它一年四季都可使用。

Mobile saunas like this in Estonia could become a gathering spot for outdoor meals.

13 Snow clearance has become an ongoing political issue for winter cities, with **disabled people, the elderly, and parents and caregivers arguing that sidewalks and crossings** [31] deserve the same priority as cars, lest people be essentially trapped in their homes. Many physically disabled people have already had their mobility limited during quarantine due to pre-existing health risks, the inability to avoid using elevators and the difficulty of maintaining social distancing. Temporary urban design changes also need to be inclusive.

14 Gabrielle Peters, a multiply disabled wheelchair user and writer, is a former member of the City of Vancouver's **Active Transportation** [32] and Policy Council. She says, "I'm a big advocate for widening sidewalks. Sidewalks shouldn't just be so that you can get from place to place. They are a place to socialize, to move at different speeds, to let a **toddler** [33] look at the grass." Among the proposals in New York City Comptroller Scott Stringer's recent report on saving the city's "Main Streets" was a call for expanding sidewalks in **commercial corridors** [34], and getting **garbage** [35] off the sidewalks. The expanded pedestrian zones would also accommodate street vendors (here's where the baked potatoes reappear), public bathrooms and seating that doesn't require a purchase.

15 "We also need large overhangs and roofed areas with heaters in public space (not just restaurant patios) where people can gather. Respiratory conditions such as **asthma** [36] are quite common and cold air can be

在爱沙尼亚，像这样的移动桑拿房可以成为户外聚餐的聚集地。

清除积雪已成为冬季城市一直面临的一个政治性问题，残疾人、老年人、为人父母者和护理人员都认为人行道和十字路口也应该与行车道一样享有优先清除积雪的权利，以免人们被困于自己家中出不了门。许多残障人士在隔离期间，因为存在潜在的健康风险、无法避免电梯的使用以及难以保持合适的社交距离，行动能力已经受到了限制。因此，临时的城市设计变更也应当将这些情况考虑在内。

Gabrielle Peters 是一名身患多种残疾的轮椅使用者，同时她也是一名作家，以及温哥华市动态交通与政策制定委员会的前成员。她说："我极力主张拓宽人行道。人行道不应该只是为了方便从一个地方走到另一个地方，而应该是社交场所，是人们可以按不同速度行走的场所，是蹒跚学步的孩子可以观察草地的场所。"纽约市审计官 Scott Stringer 在最近撰写的一份关于拯救城市主要街道的报告中，呼吁拓宽商业街的人行道并清除人行道上的垃圾。拓宽后的步行区还能容纳街头小贩（这里又出现了前文中的烤土豆）、并设有公共厕所以及可供休息的免费座椅。

"我们还需要在公共场所设置大型悬挑空间和带屋顶的区域，供人们聚集之用，并配备加热器（不局限于露天餐厅），是因为考虑到呼吸系统疾病非常常见，比如哮喘，遇到冷空气刺激就容

a trigger[37]," Peters says. "There is also the concern these changes will reduce **accessibility for disabled people**[38] depending on how they are designed and operated. There are expanded brunch options for people with funds to eat in privatized and often inaccessible public space that **encroaches on and lessens**[39] sidewalk accessibility. It seems every urbanist's favorite option is taking away **parking space**[40]. If we remove the accessible parking spaces, we remove the people who use them from public space. Some people need cars to get around."

16 Wintermission's Simor says that any **blueprint**[41] for Covid winter should be expansive enough to include families. "When we are doing engagement, children have the most positive feelings about winter. What are we doing wrong that we lose that?" he says. S'mores kits proved to be a popular way to get families to **linger**[42] in parks in the winter, at a relatively low cost, along with working with municipal fire departments to make them more comfortable with the idea of public fire pits. "A lot of public parks do have barbecue areas. If you take that same thinking, and allow fire pits, rather than only being used three or four months of the year you can use them 12 months of the year."

17 Outdoor winter activities are only as safe as any outdoor activity and, as summer exposed the impact of the unequal distribution of parks in Black and Latinx communities and the disparate effect of open streets in **low-income communities**[43], efforts to create social spaces outside have to acknowledge local conditions that may include violence, crime and drug use, as well as the over-policing of Black and Brown communities. Marc Miller, a landscape architect and professor at Penn State, recalls taking the bus in winter in the Twin Cities, and how the **bus shelters**[44] there weren't big enough to protect people with shopping carts who didn't have cars at home to go **grocery shopping**[45].

18 "It is part of the process: How do you make spaces that are accommodating that actually work for your user group, provide them with what they need to be comfortable, and then how is it not seen as a threat to other people? If there are six people over there in parkas, is that a threat?" Miller points out that even design **renderings**[46] of winter activities often lack Black scales (figures who populate architectural designs). "That creates a **gap**[47] of imagination of how Black people and people of color use spaces in the winter."

易犯病。"Peters 说，"但也有人担心这些变化可能不适合残障人士，关键问题在于如何设计和操作。有钱人有更多的选择，可以在私有化的、别人无法进入的公共场所吃早午餐，这种做法其实侵占并减少了人行道的可用空间。每个城市规划专家似乎都喜欢拿掉停车空间，但如果我们把这些可用的停车位拿掉，就相当于把使用这些停车位的人从公共空间中移除出去了，毕竟有些人必须开车出行"。

"冬季行动计划"的工作人员 Simor 说道，面对疫情，冬天的任何活动方案都应该具有足够的包容性以适用于更广泛的家庭。他说："当我们参与活动时，孩子们对冬天的反应是最积极的。我们为什么没有这份积极性？我们做错了什么？"事实证明，S'mores 工具包是一种有效的方式，使越来越多的家庭愿意到公园里去进行冬季户外活动，这种方式成本相对较低。而且，通过与市政消防部门合作，允许使用户外篝火，让人们更容易接受冬季出行的想法了。"很多公园都有烧烤区，只要你有这种想法，在允许使用篝火的情况下，你在一年 12 个月里随时都可以实现，而不仅仅是冬季的三四个月。"

冬季的户外活动是安全的，与任何户外活动一样，但需要注意的是，在努力创建户外社交空间的同时，必须对当地情况有充分认知，可能包括暴力、犯罪、吸毒情况，也可能包括黑人社区和拉丁社区的过度执法情况，因为夏天时就已经暴露出黑人社区和拉丁社区因公园分布不均产生的问题，以及低收入社区实行开放街道带来的影响。宾夕法尼亚州立大学的景观建筑师 Marc Miller 教授回忆了冬天在双子城坐公交车的情景，公交候车亭面积不够大，无法庇护那些因为家里没有车而不得不拖着购物车去购物的人。

"这只是城市空间规划的一部分：如何为使用者创造合宜的空间并提供他们所需的舒适感，而同时又不对其他人构成威胁？试想如果有 6 个穿派克大衣的人站在那里，算不算是格格不入？"Miller 指出就连冬季活动的设计效果图都很少出现黑人标度（构成建筑设计的人体形象）。他说："这就说明根本没人想到黑人和有色人种在冬天如何使用公共空间的问题。"

19 Eric Keller responded, "Nobody at Penn State is taking me up on my idea to have (masked) nightly dance parties on the student union lawn." If you can have a silent disco, why not a dance party with parkas, masks and fire pits?

20 Bring on the baked potato.

Eric Keller 回应道："宾夕法尼亚州立大学没有人会同意我每晚在学生会草坪上举办（蒙面）舞会的想法。"如果你可以开无声的迪斯科舞会，为什么不开有派克大衣、面具和篝火的舞会呢？

把烤土豆端上来吧。

（文献来源：Alexandra Lange. How to Make the Most of Covid Winter. https://www.bloomberg.com/news/articles/2020-09-11/how-to-prepare-for-a-coronavirus-winter. 杨慧翻译）

词汇 | Vocabulary

[1] make the most 极度重视，充分利用。本文讨论的是对城市公共空间的充分利用。COVID-19 即 2019 冠状病毒，因此题目 "How to Make the Most of Covid Winter" 是指如何应对疫情和寒冬的双重压力，保持城市公共空间的人气与活力

[2] street-side 街边的、道路一侧的，这里 street-side"cans" 指路边摊

[3] ubiquitous 无处不在的，随处可见的

[4] spud 马铃薯，土豆

little 表示否定意义，这里是指无论什么天气条件，卖家和买家只能站在街头，几乎没有别的方式

[5] bearable 可忍受的，不那么难受的。这里是指，正是这一点儿食物、一点儿火苗、一点儿交谈，就使得人们对寒冷冬日的户外活动不那么难以接受了。因此对城市公共空间来说可以继续保持人气了

[6] gesture （表明感情或意图的）姿态，既可以表示具体的姿势、动作，也可以表示姿态、意图。the elements 指前文提到的恶劣天气，注意与上一段的 elements 区分。文中是指烤土豆虽小，也表达了一种姿态、一种意图，是对冬天恶劣天气的一种应对

[7] portable 便携的，轻便的，可随身携带的

[8] pop-up 凸出边界的，有立体感的

　　pop-up comforts 指移动售卖车，车身可打开成为售卖窗口

[9] crowd 人群，一群人

[10] write off 唱衰。这里是指冬季还未来临，不要先行气馁

[11] *Atlantic* 此处指美国《大西洋月刊》

[12] bleak （氛围）暗淡的，凄凉的

　　unpleasant （因人导致）不愉快的

　　dismal （因事导致）阴沉的、沉闷的

[13] stretch *v.* （时间）延续、（空间）延伸、耗尽、（布料）可伸缩；*n.* （时间）连续一段、（土地）一片一段、（肢体）伸展、苛求的任务、困难的任务

stretch oneself 尽最大努力，此处联系上下文，可译为困难（苛求）的任务

[14] But now is the time to think ahead 现在是时候要提前考虑了

[15] hue 色调，文中引申为变化

[16] immune system 免疫系统

[17] 疫情也可能是这种情况，这就意味着 2020 年我们至少有两个理由要待在室外

[18] WinterCity initiative 冬城倡议，为改变埃德蒙顿人对冬季的印象，让市民 / 游客在冬季拥有更有趣的活动和节日

[19] pedway 步行立体通道系统，步行道

[20] chalk *n.* 粉笔，用粉笔画的记号；*v.* 用粉笔画

　　sidewalk chalk 人行道彩绘

[21] run 这里指商品的畅销

[22] prejudice 偏见

[23] gear up 齿轮加速，这里指为一天的生活做好准备。

[24] loiter *v.*（在公共场所）闲逛

[25] Stickiness encourages people outside 黏附力鼓励人们到来到户外

[26] pulverize 打成粉末，这里指各个社团各项活动受疫情影响完全停滞

[27] mural 壁画

　　light installation 灯光装置

[28] mayor and governor 市长和州长

[29] figure out 弄明白，清楚

[30] equivalent 同等，等效

[31] disabled people 残疾人

　　the elderly 老年人

　　caregiver 护理人员

　　sidewalk and crossing 人行道和十字路口

[32] Vancouver's Active Transportation 温哥华动态 / 积极交通

[33] toddler 初学走路的孩子

[34] commercial corridor 商业街道

[35] garbage 垃圾

[36] asthma 哮喘

[37] trigger 触发

[38] accessibility for disabled people 残障人士可得到（的可能性）

[39] encroach 侵占（土地）

    lessen 缩小，减少

[40] parking space 停车空间，停车位

[41] blueprint 蓝图，方案

[42] linger 逗留，徘徊，结合上下文，指游憩、休息

[43] low-income communities 低收入社区

[44] bus shelter 公交候车亭

[45] grocery shopping 食品杂货店

[46] rendering 渲染，效果图

[47] gap 间隙，缺口，在文化领域多有"隔阂"之意

练习与思考 | Comprehension Exercise

1. 第 10 段："Many physically disabled people have already had their mobility limited during quarantine due to pre-existing health risks, the inability to avoid using elevators and the difficulty of maintaining social distancing. Temporary urban design changes also need to be **inclusive**." 请理解该段落的意义，同时阐述你对包容性的理解以及城市该如何实现包容性。

2. 理解第 16 段中重点词语的涵义，并查阅资料了解相关背景。

Outdoor winter activities are only as safe as any outdoor activity and, as summer exposed the impact of the unequal distribution of parks in **Black and Latinx communities** and the disparate effect of open streets in **low-income communities**, efforts to create social spaces outside have to acknowledge local conditions that may include violence, crime and drug use, as well as the over-policing of Black and Brown communities.

课后延伸 | Reading Material

建设强大和包容的社区 | Building Strong and Inclusive Communities

为社区优化赋能 | Enabling Improved Neighborhoods

Building Strong and Inclusive Communities

1   London is made up of diverse communities. Its neighbourhoods, schools, workplaces, parks, community centres and all the other times and places Londoners come together give the city its cultural character and create its future. Planning for Good Growth means planning with these communities – both existing and new – making new connections and eroding inequalities.

2   London is one of the most diverse cities in the world, a place where everyone is welcome. 40 per cent of Londoners were born outside of the UK, and over 300 languages are spoken here. 40 per cent of Londoners are from Black, Asian and Minority Ethnic (BAME) backgrounds, and the city is home to a million EU citizens, 1.2 million disabled people, and up to 900,000 people who identify as LGBT+. Over a fifth of London's population is under 16, but over the coming decades the number of Londoners aged 65 or over is projected to increase by 90 per cent. This diversity is essential to the success of London's communities. To maintain this London must remain open, inclusive and allow everyone to share in and contribute towards the city's success.

3   London is one of the richest cities in the world, but it is also home to some of the poorest communities in the country, with wealth unevenly distributed across the population and through different parts of the city. It is home to an ageing population, with more and more people facing the barriers that already prevent many from participating fully in their communities. Traffic dominates too many streets across the city, dividing communities and limiting the interactions that take place in neighbourhoods and town centres.

4   Delivering good quality, affordable homes, better public transport connectivity, accessible and welcoming public space, a range of workspaces in accessible locations, built forms that work with local heritage and identity, and social, physical and environmental infrastructure that meets London's diverse needs is essential if London is to maintain and develop strong and inclusive communities.

5   Early engagement with local people leads to better planning proposals, with Neighbourhood Plans providing

建设强大和包容的社区

伦敦是由形形色色的社区组成的。邻里、学校、工作场所、公园、社区中心以及伦敦人聚集在一起的其他所有地点，都赋予了这座城市以文化特征并创造了城市的未来。为"美好增长"而规划，就意味着与这些社区（包括既有和新建社区）一起努力建立新的连接，消除不平等。

伦敦是世界上最多元的城市之一，欢迎所有人。40% 的伦敦人出生在英国以外，有 300 多种语言在这里同时使用。40% 的伦敦人有黑人、亚裔和少数族裔背景，同时这座城市也是 100 万欧盟市民，120 万残障人士，以及多达 90 万的 LGBT+ 人群的家。16 岁以下人口超过五分之一，但在未来几十年内 65 岁及以上的人口数量预计增加 90%。人口的多样性对伦敦社区的成功至关重要。为了保持这种多样性，伦敦必须保持开放、包容的态度，让每一个人都能分享这座城市的成功并为之努力。

伦敦是世界上最富有的城市之一，但是由于财富在人群和城市不同区域的分配并不均匀，英国最贫穷的社区也在伦敦。伦敦是老龄人口的家园，有越来越多的人面临出行障碍，不能充分参与社区活动。机动车在街道上占据着主导地位，分隔了社区，也限制了邻里之间与城镇之间的互动。

伦敦要保持并发展强大和包容的社区，提供可负担的高品质住房，更便捷的公共交通网络，便利舒适的公共空间，易到达的工作地点，与当地遗产和特质相适应的建筑形式，以及满足伦敦多元需求的社会、物质和环境基础设施，都是至关重要的。

早期让当地居民参与进来可以带来更好的规划建议，比如邻里计划（Neighbourhood Plans），

a particularly good opportunity for communities to shape growth in their areas. Taking advantage of the knowledge and experience of local people will help to shape London's growth, creating a thriving city that works better for all Londoners.

Enabling Improved Neighborhoods

1   Why are so few cities, towns and neighborhoods in the United States walkable? Why is it so difficult to find vibrant communities where people of all ages, incomes and backgrounds can live, work, shop and play?

2   The answer, in many locations, is that zoning codes and land use ordinances have made the creation of such places illegal. In some communities, the lack of walkability, opportunity and livability stems from zoning and development decisions that intentionally separated people by race, faith, ethnicity or income.

3   There are 42,000 units of local government with zoning authority in the United States. This guide explains why a community may want to change its zoning codes and rules, and how it can do so in ways that strengthen the local economy, promote equity, and support diversity and inclusion.

4   The Congress for the New Urbanism—with support from AARP and other partners—launched "The Project for Code Reform" to support communities that want to revise their zoning codes but don't have the staffing and resources to seek full-fledged change. For these places, the wisest path is often to pursue incremental changes that can nonetheless improve the economy, built environment and residents' quality of life.

为社区提供了促进区域增长的大好机会。充分利用本地居民的知识和经验将有助于促进伦敦的发展，创造一个欣欣向荣的城市，更好地为所有伦敦人服务。

（文章来源：Mayor of London. The London Plan 2021. https://www.london.gov.uk/what-we-do/planning/london-plan/new-london-plan/london-plan-2021. 杨慧翻译）

为社区优化赋能

为什么美国的城市、乡镇以及社区，几乎没有实现步行友好的呢？为什么很难找到一个充满活力的社区，让不同年龄、收入和背景的人都能在此愉快地生活、工作、购物和娱乐呢？

答案就是，在很多城市，根据区划法和土地使用条例的规定，建立这样的场所是不合法的。一些社区缺乏步行条件、缺少发展机会、甚至缺失宜居性，正是由于区划法与土地开发条例有意将人们分开，或按照种族、信仰、族裔，或依据收入进行区分。

在美国，有 42000 个地方政府拥有分区管理权。这本手册解释了为什么有些社区想要修改其区划法规与条例，以及如何以加强当地经济发展、促进社会公平、支持城市多样性和包容性为前提来实现。

在美国退休人员协会（AARP）和其他合作伙伴的支持下，新城市主义协会（CNU）发起了一个"法规改革项目"，用以支持那些希望修改其土地区划，但没有足够的人力和资源寻求全面变革的社区。对这些社区来说，最明智的办法往往是渐进式的变革。渐进式的变革依然能够改善经济、建筑环境和居民的生活质量。

5 *Enabling Better Places: a Handbook for Improved Neighborhoods* provides options for communities to consider as they identify and select small-scale, incremental policy changes that can be made without overhauling entire zoning codes and land use policies.

6 This handbook collaboration by AARP and CNU has been created as a reference for discussions among local leaders and community members interested in improving where they live. It is based on work led by the Michigan Economic Development Corporation and Michigan Municipal League that sought to identify incremental zoning code changes to spur economic growth. The publication is not a comprehensive checklist, nor is it meant to be used in lieu of a careful, context-specific code review process to determine and prioritize the best opportunities for beneficial change.

《营造更好的场所：社区改善手册》为社区确定并采用小规模渐进式变革政策提供了可选项，而这些小规模渐进式变革无须彻底修订整个区划法和土地使用政策。

这本由美国退休人员协会（AARP）和美国新城市主义协会（CNU）联合编写的手册，已经成为当地领导人以及对改善生活环境感兴趣的社区居民的重要参考。它是根据密歇根经济开发公司和密歇根市政联盟的工作来编写的，这些工作旨在逐步调整区划法以刺激经济增长。该手册不是一份面面俱到的清单，也无意取代细致严谨、符合特定情境的评审程序来确定并优先使用最有可能带来有益变革的区划法规。

（文章来源：AARP & CNU. A Handbook for Improved Neighborhoods. 杨慧、李善文、段语嫣、李佳琪、于露、冷雨桐，王皓天、于潇滨、刘亚杰、陈昱霖等翻译）

附录 1　城乡规划术语

Technical Terms of Urban Planning

城乡规划分析与优化技术

1. 地理信息系统（geographical information system，GIS），在计算机软、硬件支持下，对空间数据进行采集、编码、处理、存贮、分析、输出的人机交互信息系统。

2. 地理数据库（geographical database），运用计算机，对自然要素、经济要素、社会要素等空间数据进行科学组织和管理的数据库系统。

3. 网络地理信息系统（web geographic information system），地理信息系统与互联网有机结合，功能扩展，形成的分布式、超媒体特性、互操作的计算机技术系统。

4. 数字地理空间数据框架（digital geospatial data framework），基础地理信息资源及其采集、加工、分发、服务所涉及的政策、法规、标准、技术、设施、机制和人力资源的总称。

5. 数字化（digitizing），将模拟曲线转换成离散数字坐标的过程。

6. 图解法（iconography），用以解释视觉意象象征含义的描述和说明。

7. 空间网络分析（spatial network analysis），地理信息系统中依据网络拓扑关系，揭示要素空间关系的分析方法，包括最短路径分析、资源分配分析、等时性分析等。

8. 缓冲区分析（buffer analysis），地理信息系统中以点、线、面实体为基础，以距离为条件，自动生成实体周边范围，是解决邻近度问题的分析工具。

9. 叠置分析（overlay analysis），不同主题的数据层进行逻辑交、逻辑并、逻辑差等运算，生成一个新数据层，是提取空间隐含信息的分析工具。

10. 三维城市模型（3D city model），城市三维空间的数字化形式，形成城市空间的可视化数字模型。

11. 虚拟现实（virtual reality，VR），运用计算机技术，生成多源信息融合的交互式三维动态视景和模拟实体行为的仿真系统。

12. 城市仿真（urban simulation），运用虚拟现实技术模拟真实城市，实现人机交互、空间感知、三维仿真等特性的系统化方法。

13. 城市基础部门模型（urban export-base sector model），用于分析城市经济增长的经济模型。以出口（对区外服务）为基础的产业集合构成城市基础部门，为当地居民提供日常服务的部门构成非基础部门，而城市经济增长取决于基础部门和非基础部门的比例，这一比例越高，则城市经济增长率就越高。

14. 供给基础模型（the supply-base model），强调以城市内部资源要素、技术积累、专业化协作程度等供给基础来决定城市经济增长的城市模型。

15. 需求指向模型（the demand-orientated model），认为城市经济增长取决于基础部门和非基础部门的比例，比例越高城市经济增长率就越高的城市模型。

16. 投入产出模型（input-output model），从投入与产出的相互依存关系角度来研究经济系统各个部分间表现的数量方法。

17. 规划环境影响评价（environmental impact assessment on planning），分析、预测和评估规划实施后可能造成的环境影响，提出预防或者减轻不良环境影响的对策和措施，进行跟踪监测，以及与其相应的技术方法与政策制度。

18. 环境容量（environmental capacity），某一环境区域或一个生态系统在维持再生能力、适应能力和更新能力的前提下，所能容纳有机体数量的最大限度或承受的污染物最大负荷量。

19. 承载力分析（bearing capacity analysis），某一区域的资源环境本底条件对人口增长和经济发展的承载能力。

20. 适宜性分析（suitability analysis），以特定土地利用目标为导向，确定与该土地利用相关的适宜因素及其等级，分析土地对该种利用的适宜程度与制约因素，从而寻求最佳的土地利用方式和合理的规划方案。

21. 敏感性分析（sensitivity analysis），选定对项目具有影响的不确定性因素，分析、测算其影响程度和敏感性程度，进而判断项目承受风险能力的一种不确定性分析方法。

22. 城市热岛效应（urban heat island effect），由于大量人工发热、建筑物和道路等高蓄热体及绿地减少等因素影响，导致城市整体或局部地区的温度显著高于周边地区的现象。

23. 地表温度模拟（ground surface temperature simulation），设定地表气候参数，考虑地表材料、构造方式、建筑布局及树荫遮挡等因素，运用数值模拟算法推测或预测城市地表温度的方法。

24. 城市声环境模拟（urban acoustic environment simulation），基于城市环境噪声排放及噪声传播原理，模拟城市噪声传播、影响范围的数值模拟方法。

25. 城市风环境模拟（urban wind environment simulation），对建筑物周围大气流动的动力学方程进行数值求解，模拟实际风环境的方法。

26. 情景规划（scenario planning），识别城市未来发展的确定性要素和不确定性要素，系统提出多种情景及其可能条件，从而增加规划弹性、提升应变能力、构建应对体系的一种战略分析方法。

27. 低影响开发技术（low impact development technology），在城市开发与更新过程中，尽可能减少对现有环境影响的所有技术的总称。

28. 模型模拟（model simulation），对城市系统或系统元素进行抽象和概念化的基础上，对城市空间现象与过程进行的模拟表达。

29. 空间计量模型（spatial econometric model），研究空间变量与绝对位置（格局）、相对位置（距离）之间数量关系的计量模型。

30. 空间相关（spatial correlation），某位置上的变量与其他位置上的变量在空间邻近或距离远近等方面存在的相关性。

31. 空间自相关（spatial autocorrelation），一个分布区内的某一要素，依赖于该要素在其周边区域分布的现象，包括空间正相关、空间负相关两种情形。

32. 空间滞后模型（spatial lag model），考虑周边区域的被解释变量对研究区的被解释变量的影响，以空间滞后项的形式加入计量方程，形成的空间计量模型。

33. 空间误差模型（spatial errors model），考虑周边区域的解释变量对研究区的解释变量的影响，以空间误差项的形式加入计量方程，形成的空间计量模型。

34. 离散选择模型（discrete choice model），对空间离散变量进行回归与解释的数学模型，包括二元选择模型和多元选择模型。

35. 趋势面分析模型（trend-surface analysis model），地理数据的空间分布及其区域性变化趋势的数值拟合模型。

36. 用地与交通关联模型（land use and transport interaction model），为实现交通导向与土地开发的有机结合，将土地利用模型与交通需求模型进行整合运用的综合模型。

37. 聚类分析（cluster analysis），将多属性统计样本分为相对同质的群组，进行定量分析的样本归类分组技术。

38. 单中心城市模型（monocentric model），针对一个就业岗位高度集中的商务中心城市，设定区位、收入、交通成本等前提，解析地价、资本密度和人口密度等方面空间变化规律的模型。

39. 蒂伯特模型（Tiebout model），假定居民可以自由迁移至提供更好公共服务与税收组合的社区，引起辖区竞争，促使地方公共物品供给实现均衡与效率的城市模型。

40. 城市扩张模型（city expanding model / urban growth model），假定城市系统或系统内元素的空间决策规则，模拟系统或元素扩张的数量关系、形态特征、时空演化的空间模型。

41. 劳里模型（Lowry model），研究基础部门就业、服务部门就业和家庭部门行为三者之间数量与空间分布关系的城市模型。一般认为，基础部门就业主导着服务与家庭部门的数量和空间分布。

42. 引力模型（gravity model），假定某一变量与空间属性正相关，与空间距离反相关，设置相应参数分析空间联系强度的一种数学模型。

43. 势力圈分析（sphere of influence analysis），结合空间属性与空间距离，依据某种准则计算城镇影响力空间范围的一种分析方法。

44. 空间动力学模型（spatial dynamics model），描述与模拟系统元素的空间相互作用、反馈机制、组织规则、以及空间格局动态演变的定量分析模型。

45. 累积因果模型（cumulative causation model），某一因素变化引起另一因素变化，后续变化反过来加强了前一因素的变化，导致演化过程沿最初方向发展，形成累积性循环发展趋势的模型。

46. 多目标规划（multi-objective programmming），解决多目标决策问题的线性规划方法。

47. 用地平衡表法（land balance table method），根据规范对城乡用地进行比例核算与数量调整，协调各用地类型间的数量结构关系，保障城乡用地平衡合理的方法。

48. 有效性评估（assessment of effectiveness），衡量规划实施结果与规划目标之间差距的一种规划实施评价方法。

49. 一致性评估（conformity assessment），评估城乡建设现状与规划期望状况相符程度的一种规划实施方法。

50. 成本效益分析（cost-benefit analysis），通过比较规划或项目的全部成本和效益来评估其价值的一种经济决策方法。

51. 帕累托最优理论（Pareto optimality theory），资源分配实现了公平与效率的一种理想状态。在帕累托最优状态下，不可能在无人受损的情况下通过改变资源分配，而使得至少一个人受益。

52. 参与式决策法（consultative decision-making），群体或利益相关者参与公共政策制订过程及决策程序的规划方法。

53. 规划支持系统（planning support system，PSS），集合一系列计算机软件工具，为规划师、公众、政府之间建立交互联系的一套信息技术框架，以提供交互式的、集成式的、参与式的方式来处理非常务性、非结构性的规划问题。

54. 地理空间预测模型（geospatial predictive model），通过构建融合大量空间信息，来形成统计偏好相似性的地理空间模型，生成环境的统计性特征，预测未来行为发生的可能性地图。

城乡规划生态和节能技术

55. 城市环境效应（urban environmental effect），城市各项活动给自然环境带来一定程度的积极影响和消极影响的综合效果，包括污染、生物、地学、资源、景观等方面的影响。

56. 城市噪声（urban noise），城市中超过国家规定的环境噪声排放标准，并干扰他人正常生活、工作和学习的声音。

57. 城市光污染（urban light pollution），城市中可见光、紫外与红外辐射等过量光辐射对人体健康和人类生存环境造成负面影响的现象。

58. 城市环境规划（urban environmental planning），依据生态学基本原理，在对城市一定空间范围内进行环境调查、监测、评价的基础上，预测环境的可能变化，优化环境功能区划，并调整城市产业结构、发展模式与空间布局，实现保护和改善城市环境的综合部署和具体安排。

59. 城市环境功能区（urban environmental functional area），为保护和改善城市环境，依据水、大气、声、光等环境要素的分布特点与管控要求，将城市空间划分为若干不同的功能区。

60. 水环境保护功能区（functional area of water environment），根据水域使用功能、水环境污染状况、水环境承受能力（水环境容量）、社会经济发展需要以及污染物排放总量控制等要求，划定的具有特定水环境功能的区域。

61. 城市水污染控制（urban water pollution control），为恢复和维护流域水环境质量，根据水域使用功能、行政区划、水域特征和污染源分布特点等因素，对城市中不同的水域和流域空间范围实施不同的污染物排放浓度和总量控制。其所划定的特定空间范围成为城市水污染控制单元。

62. 城镇生活垃圾无害化处理（harmless disposal of urban household garbage），通过物理、化学、生物以及热处理等方法，达到不危害人体健康、不污染环境目的的垃圾处理方法。

63. 城镇生活垃圾无害化处理率（harmless disposal rate of urban household garbage），无害化

处理的城镇生活垃圾数量占城镇生活垃圾产生总量的百分比。

64. 工业固体废物处置利用率（utilization rate of industrial solid waste disposal），在一定空间范围内，工业企业当年处置及综合利用的工业固体废物总量，占当年工业企业产生的工业固体废物总量的百分比。其中，这两个总量均包括当年已处置利用的往年产生的工业固体废弃物数量。

65. 森林覆盖率（forest coverage rate），在特定区域内，林地面积占土地总面积的百分比。

66. 生态空间（ecological space），具有自然属性，以提供生态服务或生态产品为主体功能的国土空间。包括森林、草原、湿地、河流、湖泊、滩涂、岸线、海洋、荒地、荒漠、戈壁、冰川、高山冻原、无居民海岛等。

67. 城市生态系统（urban ecosystem），由城市居民及其环境相互作用而形成的具有一定功能的人工生态系统。

68. 城市生态位（urban ecological niche），由城市为满足人类生存发展提供的各种条件而形成的与生态环境之间的功能关系，决定了城市在生态系统中的时空位置。

69. 生态足迹（ecological footprint，EF），在一个地区或国家内，可提供能够维持一定量人口生存所需要的资源和消纳这些人口所产生废物的生态再生能力的地域面积。

70. 城市生态安全（urban ecological security），人与其生产、生活环境及其赖以生存的生命支持系统之间存在着可持续的良性耦合关系。

71. 生态红线（ecological red line），又称"生态保护红线"。在生态空间范围内具有特殊重要生态功能、必须强制性严格保护的区域，通常包括具有重要水源涵养、生物多样性维护、水土保持、防风固沙、海岸生态稳定等功能的生态功能重要区域，以及水土流失、土地沙化、石漠化、盐渍化等生态环境敏感脆弱区域。

72. 城市生态承载力（urban ecological carrying capacity），城市生态系统的资源和环境可容纳社会经济活动强度和一定生活水平人口数量的能力。

73. 城市生态平衡（urban ecological balance），在一定阶段内，城市生态系统内各要素之间，保持相对稳定和良性相互关系的一种状态。

74. 生态敏感性（ecological sensitivity），当一个生态系统受到系统之外人类活动或自然变化的干扰时，其对干扰和改变的敏感程度。

75. 生态适宜度（ecological suitability），不同生态因素对给定的土地利用方式的适宜状况和程度。是土地利用开发适宜程度的依据。

76. 城市生态基础设施（urban ecological infrastructure），保障城市生态系统基本安全和良性循环的各种生态要素类型、组成及其空间分布形态与格局关系。生态要素包括动植物、土地、矿物、水体、气候等自然物质要素，以及地面、地下的各种建筑物和相关设施等人工物质要素。

77. 生态补偿机制（ecological compensation mechanism），以保护生态环境、促进人与自然和谐发展为目的，根据生态系统服务价值、生态保护成本、发展机会成本，运用政府和市场手段，调节生态保护利益相关者之间利益关系的公共制度。

78. 生态修复（ecological restoration），对遭到破坏的生态系统辅以人工措施，加速其恢复或向良性循环方向发展的行为。

79. 城市生态修复规划（urban ecological restoration planning），运用生态学原理，对城市受损生态系统进行评价、规划，综合安排加快山体修复、开展水体治理和修复、修复和利用废弃地、完善绿地系统等措施，以达成人与自然和谐、促进生态系统恢复和重建的专项规划。

80. 城市生态功能区划（urban ecological function zoning），依据区域生态特征、生态系统服务功能和生态环境敏感性的空间分异规律，将特定区域划分为不同地域单元，并确定各单元的主导生态功能。

81. 城市生态用地（urban ecological land），用于提高城市居民生活质量，保护重要的生态系统和生物栖息地，完善城市各种生态功能，维持城市生态系统稳定的用地。

82. 生态缓冲区（ecological buffer），在城市与重要生态功能区交界地带，以植物为主、发挥一定缓冲作用的过渡区域。

83. 生态廊道（ecological corridor），具有保护生物多样性、过滤污染物、防止水土流失、防风固沙、调控洪水等生态服务功能的廊道。一般以植被、水体为主。

84. 生态隔离带（ecological isolation zone），在城市外围或者组团之间，以林地、湿地、农业用地、园地等生态用地为主的绿色植被带。

85. 城市节能技术（urban energy-saving technology），城市建设领域内，为实现节约能源目的而采用的技术手段的总称。一般包括节电、节煤、节油、节水、节气、节地以及工艺改造节能技术等。

86. 城市能源规划（urban energy planning），对城市能源生产和消费状况进行调查和分析，预测能源需求，并对能源的开发、生产、转换、使用和分配而进行的统筹安排。

87. 城市能源结构（urban energy structure），在城市内能源的生产和消费过程中，各种一次能源、二次能源的构成及其比例关系。包括能源生产结构、能源消费结构。

88. 能源梯级利用（cascade utilization of energy），按照能源的品位（能源转换效率）逐级加以合理利用的用能方式，如高、中温蒸汽先用于发电或生产，低温余热用于供热。

89. 能源综合利用（comprehensive use of energy），采用新能源技术、加强能源梯级利用、促进能源节约、提高能源利用效率等，以构建安全、高效、可持续的能源利用体系。

90. 城市微气候环境（urban micro-climatic environment），城市特定区域或空间内，因人工或自然空间环境与城市所在广域空间环境的不同，形成气温、风象、降雨等要素差异化的局地气候环境。其差异也常常出现在城市内各个特定区域或空间之间。

91. 气候调节（climatic regulation），依据生物调节气候的原理，采用生态规划的技术方法，利用自然气候营造宜人的微气候环境的技术。

92. 城市风廊（urban wind corridor），以提升城市的空气流动，缓解热岛效应和改善人体舒适度为目的，为城区引入新鲜空气而构建的通道。

93. 楔形绿地（green wedges），从城市外围嵌入城市内部的连续、成片绿地。

94. 街道峡谷（street canyon），城市中由街道及两侧一系列相对连续的建筑界面形成的、类似于峡谷的街道空间形态。

95. 高层建筑风影区（wind shadow of high-rise buildings），在高层建筑的背风侧出现的与原有风向不一致的湍流区。

96. 单位国内生产总值能耗（Gross Domestics Product energy intensity），简称"单位 GDP 能耗"。一次能源供应总量与国内生产总值（GDP）的比率。是反映能源消费水平和能源利用效率的指标。

97. 碳足迹（carbon footprint），直接或间接支持人类活动所产生的二氧化碳及其他温室气体总量，通常用产生的二氧化碳吨数来表示。

98. 温室气体（greenhouse gas），大气中能够吸收地面反射的太阳辐射并重新发射辐射，使地球表面变暖的气体。如二氧化碳、甲烷、水蒸气等。

99. 温室效应（greenhouse effect），太阳短波辐射可以透过大气射入地面，而地面增暖后放出的长波辐射却被大气中的二氧化碳等物质所吸收，从而产生大气变暖的效应。

100. 低碳经济（low-carbon economy），通过技术创新、制度创新、产业转型、新能源开发等多种手段，尽可能地减少传统化石能源等高碳能源消耗，减少温室气体排放的经济发展模式。

101. 环保产业（environmental protection industry），在国民经济结构中以防治环境污染、改善生态环境、保护自然资源为目的所进行的技术开发、产品生产、商业流通、资源利用、信息服务、工程承包、自然保护等活动的总称。

102. 碳贸易（carbon trade），又称"碳排放交易"。以促进全球温室气体减排为目的，把二氧化碳排放权作为一种商品形成的二氧化碳排放权的交易。

103. 城市碳排放（urban carbon emission），城市生产、生活活动过程中所有含碳元素的气体排放总量。

104. 城市碳平衡（urban carbon balance），城市在碳的排放和吸收两方面相等或相抵。

105. 低碳交通（low-carbon transportation），以高能效、低能耗、低污染、低排放为特征的交通运输发展方式，以减少传统化石能源等高碳能源的消耗的状态。

106. 慢城（slow city），慢节奏的城市生活模式及其城市空间形态，一般为人口 5 万人以下城镇或社区，减少机动交通，增加绿地和徒步区域，保护城市的个性特色和本地象征性产品，没有快餐区和大型超市。其实质也是一种放慢节奏的生活方式，包括慢餐、慢行、慢学校等。

107. 循环产业园区（circular industrial park），在生产过程和生产资料流通等方面，实现减量化、再利用、资源化经济活动的产业园区。

108. 静脉产业园区（venous industrial park），将生产和消费过程中产生的废物转化为可重新利用的资源和产品，实现再利用、资源化和环境安全的产业园区。

109. 低碳产业园区（low-carbon industrial park），在建设、生产、管理等全过程中，统筹兼顾碳排放与可持续发展，实现碳排放量最小化的产业园区。

110. 单位 GDP 碳排放（carbon emission per unit of GDP），单位国内生产总值的二氧化碳排放量。一般以万元国内生产总值（GDP）排放的二氧化碳数量统计。

城乡规划智能技术与方法

111. 智慧社区（smart community），利用物联网、云计算、移动互联网、智能交通等信息技术，提供智能化社会管理与服务的一种新的社区形态，能够使居民的生活更加便捷、舒适、高效、安全。

112. 智能交通（intelligent transportation system，ITS），利用信息通信技术对交通信息进行收集、处理、发布、交换、分析和利用，使交通运行更加智能、安全、节能、高效，并为交通参与者提供多样化服务的交通管理模式。

113. 智慧城市规划（smart city planning），充分借助物联网、云计算、规划支持系统等技术，通过多知识融合与数据挖掘等技术，以实现城市发展智能决策、智能运行等的一系列规划。

114. 数字城市（digital city），综合运用地理信息系统、多媒体、虚拟现实等数字技术，以传感方式自动采集城市数据，以数字化方式展现城市多元信息，并进行动态监测管理，最终为城市管理、生产、生活等提供服务的虚拟城市形态。

115. 数字城市规划（digital urban planning），以数字化的手段来实现对城市空间资源的有效配置与合理安排的一种规划方式。

116. 城市信息系统（urban information system），在计算机软硬件支持下，把各种与城市有关的信息按照空间分布及属性，以一定的格式输入、处理、管理、分析、输出的计算机技术系统。是用以反映城市规模、生产、生活、功能结构、生态环境及其管理的信息系统。

117. 规划信息系统（planning information system），进行城市规划的各种信息系统的统称。是计算机、通信、网络、地理信息系统、遥感、数据库等技术在城市规划与管理中的综合应用。

118. 规划管理信息系统（information system of urban planning administration），应用信息系统技术支撑规划行政许可工作的信息平台，是覆盖城市规划实施管理全过程的图文一体化办公自动化系统。

119. 行政审批一体化平台（integrated administrative approval platform），运用信息网络技术，通过信息化管理和行政审批数据共享，来提高行政审批工作的规范化、科学化和透明度的一种信息化平台。

120. 规划"一张图"（one-map of plans），以现状信息为基础，以法定图则为核心，系统整合各类规划成果，具备动态更新机制的规划信息综合管理和服务平台。

121. 时空数据挖掘（spatio-temporal datamining），从海量、高维的时空数据中提取出隐含的、人们事先不知道的、但又潜在有用的信息和知识的过程。包括时空分布规律、时空关联规则、时空聚类规则、时空特征规则和时空演变规则等。

122. 大数据（big data），一种极为巨大复杂的数据形式，具有海量的数据规模、快速的数据流转、多样的数据类型等特征，传统的数据处理或管理方法无法应用在这类数据上。

123. 物联网（Internet of things，IoT），（1）狭义上的物联网指连接物品和物品的网络，实现物品的智能化识别和管理。（2）广义上的物联网则是信息空间与物理空间的融合，将一切事物数字化、网络化，在物品之间、物品与人之间、人与现实环境之间实现高效信息交互的网络。

124. 信息与通信技术（information and communication technology，ICT），信息技术与通信技

术相融合而形成的一个新的概念和新的技术领域。

125. 规划数据库（planning database），根据城市规划编制和城市规划管理来设计的，专门用来存放与城市规划有关数据和信息的数据库。

126. 规划数据标准（standard for planning data），针对城市规划数据制定的统一的数据框架和标准，包括统一的数据分类、分级、记录格式、编码、质量等，可以规范规划编制成果和规划信息系统中的数据内容，推动城市规划信息共享。

127. 3S 技术（3S technology），地理信息系统（GIS）、遥感（RS）、全球定位系统（GPS）的统称和集成。

128. 空间数据（spatial data），用来表示空间实体的位置、形状、大小、分布特征、相关属性等方面信息的数据。包括点、线、面以及实体等基本空间数据结构。

129. 空间数据模型（spatial data model），地理信息系统（GIS）中组织空间数据的模型。是对现实世界的抽象表达。

130. 决策支持系统（decision support system，DSS），辅助决策者通过数据、模型和知识进行决策的计算机应用系统。其为决策者提供了分析问题、建立模型、模拟决策过程和制定方案的环境，可以帮助决策者提高决策水平和质量。

131. 空间决策支持系统（spatial decision support system），面向空间问题领域的决策支持系统，主要用于求解难于具体描述和模拟的空间问题。

132. 元胞自动机（cellular automata，CA），以元胞空间（例如城市网格）中的元胞（例如城市网格中的一个地块）作为基本单元，以元胞当前状态（例如用地性质）及其邻居状态确定下一时刻该元胞状态作为基本规则，通过每个元胞的局部动态变化来仿真大规模复杂问题（例如城市用地演变）的一种动力学模型。

133. 多智能体系统（muti-agent system，MAS），由超过一个智能体组成的联合决策系统，系统中的每一个智能体都是一个独立的决策个体（例如一个家庭），它通过自己从环境中获得的信息自主地作出决策（例如家庭选址），并和其他智能体进行协作、协商、交流，从而仿真大规模复杂问题（例如城市住区分布）。

（文献来源：城乡规划学名词审定委员会. 城乡规划学名词. 北京：科学出版社，2021）

附录 2　可持续城市空间与设计

Sustainable Urban Space and Design

1. 温室气体（greenhouse gases，GHGs），指大气中由自然或人为产生的，能够吸收和释放地球表面、大气本身和云所发射的陆地辐射谱段特定波长辐射的气体成分。该特性可导致温室效应。水汽（H_2O）、二氧化碳（CO_2）、氧化亚氮（N_2O）、甲烷（CH_4）和臭氧（O_3）是地球大气中主要的 GHG。此外，大气中还有许多完全由人为因素产生的 GHG，如《蒙特利尔协议》所涉及的卤烃和其他含氯和含溴物。除 CO_2、N_2O 和 CH_4 外，《京都议定书》还将六氟化硫（SF_6）、氢氟碳化物（HFC）和全氟化碳（PFC）定义为 GHGs。

2. 温室效应（greenhouse effect），大气中所有红外线吸收成分的红外辐射效应。温室气体（GHGs）、云和少量气溶胶吸收地球表面和大气中其他地方放射的陆地辐射。这些物质向四处放射红外辐射，但在其他条件相同时，放射到太空的净辐射量一般小于没有吸收物情况下的辐射量，这是因为对流层的温度随着高度的升高而降低，辐射也随之减弱。GHG 浓度越高，温室效应越强，其中的差值有时称作强化温室效应。人为排放导致的 GHG 浓度变化可加大瞬时辐射强迫。作为对该强迫的响应，地表温度和对流层温度会出现上升，就此逐步恢复大气顶层的辐射平衡。

3. 气候变化（climate change），指气候平均状态统计学意义上的巨大改变或者持续较长一段时间（典型的为 30 年或更长）的气候变动。气候变化不但包括平均值的变化，也包括变率的变化。《联合国气候变化框架公约》将其定义为：经过相当一段时间的观察，在自然气候变化之外由人类活动直接或间接地改变全球大气组成所导致的气候改变。

4. 二氧化碳当量（carbon dioxide equivalent），为统一度量整体温室效应的结果，需要一种能够比较不同温室气体排放的量度单位，由于二氧化碳增温效应的贡献最大，因此，规定二氧化碳当量为度量温室效应的基本单位，用作比较不同温室气体排放的量度单位。通过全球增温潜势进行换算。

5. 全球增温潜势（global warming potential，GWP）：在一定时期（通常为 100 年）内，排放到大气中的 1 千克温室气体的辐射强迫与 1 千克二氧化碳的辐射强迫的比值。

6. 碳强度（carbon intensity），按另一个变量（如国内生产总值、产出能源的使用或交通运输等）单位排放的二氧化碳量。

7. 碳截留（carbon sequestration），增加除大气之外碳库的碳储量的过程。

8. 碳排放（carbon emissions），指煤炭、天然气、石油等化石能源燃烧活动和工业生产过程以及土地利用、土地利用变化与林业活动产生的温室气体向大气的排放，以及因使用外购的电力和热力等所导致的间接温室气体向大气的排放。

9. 人为排放（anthropogenic emissions），人类活动引起的各种温室气体、气溶胶，以及温室气体或气溶胶的前体物的排放。这些活动包括各类化石燃料的燃烧、毁林、土地利用变化、畜牧业生产、化肥施用、污水管理以及工业流程等。

10. 直接排放（direct emissions），在定义明确的边界内各种活动产生的物理排放，或在某一区域、经济部门、公司或流程内产生的排放。

11. 间接排放（indirect emissions），在定义明确的范围内，如某个区域、经济部门、公司或流程的边界内各种活动的后果，但排放是在规定的边界之外产生的排放。例如，如果排放与热量利用有关，但物理排放却发生在热量用户的边界之外，或者排放与发电有关，但物理排放却发生在供电行业的边界之外，那么这些排放可描述为间接排放。

12. 适应（adaptation），针对实际的或预计的气候及其影响进行调整的过程。在人类系统中，适应有利于缓解或避免危害，或利用各种有利机会。在某些自然系统中，人类的干预也许有助于适应预计的气候及其影响。

13. 适应能力（adaptive capacity），指某个系统、机构、人类及其他生物针对潜在的损害、机遇或后果进行调整、利用和应对的能力。

14. 气溶胶（erosol），空气中悬浮的固态或液态颗粒物，其大小一般在几纳米至10微米之间，可在大气中驻留至少几个小时。气溶胶既包括颗粒物也包括悬浮的气体。气溶胶有自然的和人为的两类来源。气溶胶可通过几种方式影响气候：通过散射和吸收辐射直接影响；通过作为云凝结核或冰核，改变云的光学特性和云的生命周期而产生间接影响。无论是自然的还是人为的大气气溶胶，都起源于两种不同的路径：初级颗粒物的排放，然后从气态前体形成二级颗粒物。大部分气溶胶来源于自然。

15.《联合国气候变化框架公约》（United Nations Framework Convention on Climate Change, UNFCCC），简称《公约》。公约于1992年5月9日在纽约通过，并于1992年在里约热内卢地球峰会上由超过150个国家和欧洲共同体签署，《公约》在控制气候变化领域有着基石意义。《公约》由序言、二十六条正文和两个附件组成。包括公约目标、原则、承诺、研究与系统观测、教育培训和公众意识、缔约方会议、秘书处、公约附属机构、资金机制和提供履行公约的国家履约信息通报及公约有关的法律和技术等条款。UNFCCC 也是负责支持《公约》实施的联合国秘书处的名称，其办公室位于德国波恩。该秘书处，在政府间气候变化专门委员会（IPCC）的相助下，旨在通过会议和有关各项战略的讨论取得共识。《公约》的最终目标是"将大气中的温室气体浓度稳定在一个能使气候系统免受危险的人为干预的水平上"。在"共同但有区别的责任"原则下，《公约》包含了针对所有缔约方的承诺。《公约》中的附件约定一缔约方的共同目标是在 2000 年前将未受《蒙特利尔议定书》管控的温室气体排放量恢复到 1990 年的水平。《公约》于 1994 年 3 月开始生效。1997 年 UNFCCC 通过了《京都议定书》。

16. 附件 I 国家和非附件 I 国家（Annex I Parties and Non-Annex I Parties），根据《联合国气候变化框架公约》，附件 I 国家包含美国、日本、澳大利亚等 24 个经济合作组织（OECD）成员国，俄罗斯等 14 个经济转型国家，此外还有欧盟、摩纳哥、支敦士登。其他的缔约国则通称为非附件 I 国家，非附件 I 国家全部是发展中国家。附件 I 国家和非附件 I 国家承担部分共同的义务，包括制定关于本国温室气体排放情况的国家清单，制定减缓温室气体排放和适应全球变暖的国家计划，开发节能减排的科学技术，促进节能减排的技术交流，植树造林大力营造温室气体的吸收汇，普及公众绿色知识。

17. 政府间气候变化专门委员会（Intergovernmental Panel on Climate Change, IPCC），IPCC 由世界气象组织（WMO）和联合国环境规划署（UNEP）于 1988 年组织设立，是一个政府间科学机构，也是牵头评估气候变化的国际组织，其作用是对与人类引起的气候变化相关的科学、技术和社会经济信息进行评估，旨在提供有关气候变化的科学技术和社会经济认知状况、气候变化原因、潜在影响和应对策略的综合评估。本组织的工作与政策具有相关性，但又对政策保持着中立关系，不对政策作任何指令或规定。IPCC 由三个工作组和一个专题组组成。工作组和专题组由技术支持小组（TSU）予以协调。三个工作组分别是：第一工作组负责气候变化的自然科学基础研究，第二工作组负责气候变化的影响、适应和脆弱性研究，第三工作组负责减缓气候变化研究。国家温室气体清单专题组的主要目标是制定和细化国家温室气体排放和清除的计算和报告方法。

18.《京都议定书》（Kyoto Protocol, KP），KP 是 1997 年在日本京都召开的《联合国气候变化框架公约》第三次缔约方大会上通过的国际性公约。为发达国家的温室气体放量规定了标准，即在 2008 至 2012 年间，全球主要工业国家（附件 I 国家）的工业二氧化碳排放量比 1990 年的排放量平均要低 5.2%。

19. 可测量、可报告和可核查（"三可"原则）（measurement, reporting and verification, MRV），是指根据制定的相关温室气体核算、报告的指南或方法学，完成相应区域、机构、组织或项目的温室气体排放和清除的量，且监管或管理机构也可按相应的指南或方法学对其进行核查的原则，是国际社会、组织和机构对温室气体排放和减排核算监测与报告的基础要求。

20. 森林（forest），指最小面积 0.05 至 1.0 公顷的土地上，郁闭度（或同等存量水平）大于10%、就地树高度达到 2~5 米。森林可为由具有不同高度层次的树木和下层灌木覆盖很大部分地面的郁闭林或疏林组成。幼年天然林地和树冠密度可达到 10% 至 30% 或树高 2~5 米的所有种植园均包括在森林范围内；由于人类干扰（如采伐或自然因素）暂时无林木但可望恢复为森林的，通常看成是森林的一部分，也属森林范围。我国定义：森林是指土地面积大于等于 0.067 公顷，郁闭度大于等于 0.2，就地生长高度可达到 2 米以上（包含 2 米）的以树木为主体的生物群落，包括天然与人工幼林，符合这一标准的竹林，以及国家特别规定的灌木林，行数在 2 行以上（含 2 行）且行距小于等于 4 米或冠幅投影宽度在 10 米以上的林带。

21. 森林碳汇（forest carbon sink），指森林植物群落通过光合作用吸收大气中的二氧化碳将其固定在森林植被和土壤中的所有过程、活动或机制。

22. 林地（forest land），这一类别包括拥有与国家温室气体（GHG）清单中用来界定森林的阈值相一致的木本植被的所有土地，在国家一级细分为经营和非经营林地，并且也按《2006IPCC 国家温室气体清单指南》中规定的生态系统类型细分。它还包括其植被目前低于但可望超过林地类别阈值的系统。

23. 森林管理（forest management），林地管理和使用的做法体系，旨在实现森林生态（包括生物多样性）、经济和社会功能可持续性。

24. 森林清查（forest inventory），测量森林面积、数量和分布状况的调查系统，通常采用连续抽样调查进行。

25. 立木材积（standing volume），活立木或枯立木的带皮体积，是指自树干根基部到树梢、并大于一定胸径范围的主干带皮体积（材积），我国活立木材积测定最小起测胸径为 5.0cm。

26. 活立木蓄积量（growing stock），所有活立木材积总量（单位：立方米），包括森林、疏林、散生和四旁乔木材积。

27. 森林蓄积量（forest stock），指森林内达到检尺范围的所有立木材积总量（单位：立方米）。

28. 毁林（deforestation），人为直接引起的林地向非林地的转变。

29. 造林（afforestation），通过栽植、播种和 / 或人工促进天然更新方式，将至少 50 年以来的无林地转化为有林地 的人为直接活动。

30. 再造林（reforestation），指森林或林地经人为砍伐殆尽之后，通过自然或人为的方式，使其再次成林的过程。《京都议定书》第一个承诺期中的再造林活动，是指在 1989 年 12 月 31 日至今无森林的土地上重新恢复森林。

31. 疏林（open forest），是由树冠郁闭度大于或等于 10% 及小于 20% 的稀疏乔木植物组成的群落。

32. 经营林（managed forest），指所有人类干预和相互作用的森林（主要包括商业性管理、木材采伐和薪柴、商品木材的生产和利用以及为实现国家规定的景观或环境保护而管理的森林），具有确定的地理边界。

33. 库（或碳库）（pool（carbon pool）），具有累积或释放碳的能力的库或系统，碳库的实例有森林生物量、土壤和大气层。森林碳库通常包括地上生物量、地下生物量、枯落物、枯死木和土壤有机质五个碳库，其单位为质量单位。此外，木质林产品也可以视作是一个碳库，单位是质量单位。

34. 生物量（biomass），生态系统中植物地上、地下、活的和枯死的有机干物质，例如树木、作物、草及其枝叶、根等。生物量包括地上和地下生物量。

35. 干物质（dry matter, d.m.），指已经烘干后的有机物。

36. 地上生物量（above-ground biomass），土壤层以上以干重表示的植被所有活体的生物量，包括干、桩、枝、皮、种子、花、果和叶及草本植物。

37. 地下生物量（below-ground biomass），所有活根生物量，通常不包括难以从土壤有机成分或枯落物中区分出来的细根（直径 ≤ 2.0 mm）。

38. 枯死木（dead wood），枯落物以外的所有死生物量，包括枯立木、枯倒木以及直径 ≥ 5.0 cm 的枯枝、死根和树桩。

39. 枯落物（litter），土壤层以上、直径小于 ≤ 5.0 cm、处于不同分解状态的所有死生物量。包括凋落物、腐殖质，以及难以从地下生物量中区分出来的细根。

40. 低活性黏土土壤（low activity clay（LAC）soil），含有低活性黏土（LAC）矿物质的土壤为高度风化的土壤，以 1 1 的黏土矿物质和非晶态氧化铁及氧化铝为主（粮农组织分类中包括：淋溶土、强风化弱粘淀土、铁铝土）。

41. 高活性黏土土壤（high activity clay（HAC）soil），含有高活性黏土矿物质的土壤是轻度至中度风化的土壤，硅化黏土矿物质比例为 2∶1（粮农组织分类包括：变性土、黑钙土、黑土、淋溶土）。

42. 腐殖质层（humus horizon），该层主要由呈细粒分布的有机物质组成（但仍在矿质土层的上层）。肉眼可辨的植物残余部分依然存在，但数量比细粒分布的有机物质少得多。该层可含有矿质土壤颗粒。

43. 土壤有机碳（soil organic carbon），一定深度内（通常为 1.0 m）矿质土和有机土（包括泥炭土）中的有机碳，包括难以从地下生物量中区分出来的细根（小于 2 mm）。

44. 收获木质林产品（harvest wood products，HWP），包含木材纤维类产品和部分非木材纤维的竹藤类产品。这里主要是指以木质材料为原料 加工的各类产品，包括圆木、工业圆木、薪材（包括木炭）、锯木、木板、其他工业原木（产品）、纸浆、纸和纸板以及回收纸等木材纤维产品。

45. 基本木材密度（basic wood density），烘干树干重量与新鲜树干体积（不包括树皮）的比值。它是以干物质质量计算的木材生物量。

46. 生物量扩展系数（biomass expansion factor，BEF），树木地上生物量与树干生物量的比值。

47. 生物量换算和扩展系数（biomass conversion and expansion factor，BCEF），树木地上生物量与树干材积之比，单位：吨干物质 / 米^3。

48. 碳循环（carbon cycle），碳循环是一种生物地质化学循环，指碳元素在地球上的生物圈、岩石圈、水圈及大气中 交换。碳的主要来源有四个，分别是大气、陆上的生物圈（包括淡水系统及无生命的有机化合物）、海洋及沉积物。通过化学、物理和生物过程进行从库到库的碳交换。碳循环与氮循环和水循环一起，包含了一系列使地球能持续存在生命的关键过程和事件。碳循环描述了碳 元素在地球上的回收和重复利用，包括碳沉淀。

49. 碳密度（carbon density），单位面积的碳储量，通常指有机碳。

50. 碳通量（carbon flux），指碳循环过程中，在单位时间单位面积二氧化碳从一个库向另一个库的转移量。

51. 碳中和（carbon neutral），也称碳补偿（carbon offset）。"碳中和"是指通过计算某活动、工业生产或其他相关活 动导致的二氧化碳排放总量，然后通过造林、森林经营等碳汇项目产生的碳汇量（减排量）抵 消了相应的排放量，以实现碳排放与碳清除相互抵消，达到中和的目的。

52. 碳储量（carbon stock），一个库中碳的数量，单位：吨碳（tC）。

53. 碳储量变化（carbon stock change），碳库中的碳储量由于碳增加与碳损失之间的差别而发生的变化。当损失大于增加时，碳储量变小，因而该碳库为源；当损失小于增加时，该碳库为汇。

54. 汇（sink），从大气中清除温室气体、气溶胶或温室气体前体的任何过程、活动或机制。

55. 源（source），向大气中排放温室气体、气溶胶或温室气体前体的任何过程或活动。

56. 排放系数（emission factor），表述一种空气污染物在某种活动下的产生比例，由污染物产生量与活动的比值计算出来。在给定运行条件下某一类活动的排放系数通常是基于测量数据的样本得到的平均代表性排放率。

57. 燃烧效率（combustion efficiency），以二氧化碳形式释放的燃烧碳的比例。

58. 碳交易（carbon trading），《京都议定书》为促进全球减少温室气体排放，以国际公法作为依据的温室气体减排量交易，即是温室气体二氧化碳排放权交易。在 6 种被要求减排的温室气体中，二氧化碳（CO_2）为最大宗，所以这种交易以每吨二氧化碳当量（tCO_{2e}）为计算单位，通称为"碳交易"。其交易市场称为碳市（carbon market）。

59. 碳交易机制（carbon trading mechanism），碳交易机制是规范国际碳交易市场的一种制度。碳资产原本并非商品，也没有显著开发价值。1997 年《京都议定书》的签订改变了这一切。按照《京都议定书》的规定，到 2010 年所有发达国家排放的二氧化碳、甲烷等在内的 6 种温室气体数量要比 1990 年减少 5.2%。但由于发达国家能源利用效率高，能源结构优化，新能源技术被大量采用，因此本国进一步减排的成本高，难度较大。而发展中国家能源效率低，减排空间大，成本也低。这导致同一减排量在不同国家之间存在不同成本，形成价格差。发达国家有需求，发展中国家有供应能力，碳交易市场便由此产生。为达到《联合国气候变化框架公约》全球温室气体减量的最终目的，依据公约的法律架构，《京都议定书》中规定了三种减排机制：清洁发展机制（clean development mechanism，CDM）、联合履约（joint implementation，JI）和排放贸易（emissions trade，ET）。

60. 清洁发展机制（clean development mechanism, CDM），《京都议定书》中引入的灵活履约的机制之一。它允许缔约方与非缔约方联合开展二氧化碳等温室气体减排项目。这些项目产生的减排数额可以被缔约方作为履行他们所承诺的限排或减排量。

61. 排放贸易（emissions trade，ET），是《京都议定书》中引入的灵活履约的机制之一。它是在附件一国家的国家登记处（national registry）之间，进行包括"减排量单位"（emission reduction unit，ERU）、"核证减排量"（certified emission reductions，CERs）、"分配数量单位"

（assigned amount unit，AAU）、"清除单位"（removal unit，RMU）等减排单位核证的转让或获得。也就是发达国家将其超额完成的减排义务指标，以贸易方式直接转让给另外一个未能完成减排义务的发达国家。

62. 联合履约（joint implementation，JI），是《京都议定书》中引入的灵活履约机制之一。是附件一国家之间在"监督委员会"（supervisory committee）监督下，进行减排量单位核证与转让或获得，所使用的减排单位为"减排量单位"（ERU）。作为可交易的商品，ERU 可以帮助附件一国家实现京都议定书下的减排承诺。

63. 碳排放权（carbon emission right），指依法取得的向大气排放温室气体的权利。

64. 排放配额（emission allowances），是政府分配给重点排放单位指定时期内的碳排放额度，是碳排放权的凭证和载体。1 单位配额相当于 1 吨二氧化碳当量。

65. 额外性（additionality），指拟议的减缓项目、减缓政策或气候融资的减排项目活动所产生的项目减排量高于基线减排量的情形。这种额外的减排量在没有拟议的减排项目活动时是不会产生的。林业碳汇项目的额外性是指碳汇量高于基线碳汇量的情形，并且这种额外的碳汇量在没有碳汇造林项目活动时是不会产生的。

66. 泄漏（leakage），由于减排项目活动引起的、发生在项目活动边界外的、可测定的温室气体源排放的增加量。泄漏还指在某块土地上进行的无意识的固碳活动（例如植树造林）直接或间接地引发了某种活动，该活动可以部分或全部抵消最初行动的碳效应。无论是一个项目、县、州、省、国家，还是世界上的某个区域，每个层面都可能发生泄漏现象。

67. 核算（accounting），将报告的排放量和清除量与承诺量进行比较并按国际规则或方法学进行的相关调查和计算。

68. 碳预算（carbon budget），碳库间或碳循环的某个具体环圈（例如大气层—生物圈）间碳交换的平衡。碳库预算的审查提供了判断是源或汇的信息。

69. 碳信用（carbon credit），国际有关机构依据《京都议定书》等国际公约，发给温室气体减排国、用于进行碳贸易的凭证。一个单位的碳信用通常等于吨或相当于 1 吨二氧化碳的减排量。

70. 碳信用（carbon credit），国际有关机构依据《京都议定书》等国际公约，发给温室气体减排国、用于进行碳贸易的凭证。一个单位的碳信用通常等于吨或相当于 1 吨二氧化碳的减排量。清洁技术的推广应用会得到额外的补偿，因此这对清洁技术的研发和使用起到激励作用。在很多用于评估减缓经济成本的模型中，碳价通常被用来作为表示减缓政策努力程度的替代参数。

71. 碳金融（carbon finance），由《京都议定书》而兴起的低碳经济投融资活动，或称碳融资和碳物质的买卖，即服务于限制温室气体排放等技术和项目的直接投融资、碳排放权交易和银行贷款等金融活动。

72. 基线（baseline），用于衡量变化大小的基准数据。项目活动的基线是指在没有拟议的项目活动中人为温室气体排放量的合理预期。

73. 基线情景（baseline scenario），指在没有拟议的项目活动时，项目边界内的活动的未来情景。

74. 打捆（bundle），将几个小型碳交易机制项目活动放在一起，作为一个项目活动或活动组合，而不失各个项目活动的具体特征。这些特征包括：技术或措施、位置、简化基准线、方法学应用。每个更小的捆内的各个项目活动属于同一类型，每个小捆中的项目的产出能力不得超过相关产出类型的小规模上限。

75. 拆分（debundle/split），将一个大型项目活动拆分成多个小型项目活动。大型项目活动或大型项目活动的任何组成部分，都能采用正常的碳交易机制程式和程序。

76. 活动规划（programme of activities，POA），指为执行政府政策/措施或者实现规定的目标（例如物质激励制度和自愿项目），由私人或者公共实体自愿参与协调并执行的活动。在某一规划方案之下，可以通过添加不限数量的相关碳交易机制规划活动使之与没有此规划方案活动的情景相比，产生额外的温室气体减排或者增加温室气体汇的效益。

77. 部分项目活动（component project activity，CPA），在基准线方法定义指定的区域内实施的一个或者一系列相互关联的减排或者增汇措施。

78. 项目参与方（project participant），项目参与方是指就如何分配所考虑的项目活动产生的经核证的减排量（CERs）所作出决定的缔约方或私营和/或公共实体。

79. 项目活动（project activity），指一项旨在减少温室气体排放量的措施、操作或行动。

80. 项目边界（project boundary），指由对拟议项目所在区域的林地拥有所有权或使用权的项目参与方（项目业主）实施森林经营碳汇项目活动的地理范围。一个项目活动可在若干个不同的地块上进行，但每个地块应有特定的地理边界，该边界不包括位于两个或多个地块之间的林地。项目边界包括事前项目边界和事后项目边界。

81. 项目情景（project scenario），指拟议的项目活动下，对 GHG 排放趋势情景的预测。

82. 利益相关方（stakeholder），指受到或可能受到所拟议的清洁发展机制项目活动或导致实施此种活动的行动影响的公众，包括个人、群体或社区。

83. 指定政府主管部门（designated national authority，DNA），缔约方要想参加 CDM 项目，需要设立负责监管 CDM 的指定国家主管机构。有关缔约方需要向 UNFCCC 秘书处提供有关其 DNA 的信息。这些信息可以从 UNFCCC 的网站上查询。指定国家主管机构（DNA）是指依照国内法律和政策以及国际 CDM 规则，负责对 CDM 项目实行批准 程序的政府部门。在我国，指定国家主管机构是国家发展和改革委员会（NDRC），其负责对国家 CDM 理事会审核并修订后的项目进行批准程序。

84. 指定经营实体（designated operational entity，DOE），指定经营实体（DOE）在碳交易机制程序中扮演着非常重要的角色。它是负责请求和实施碳交易机制项目活动的合格性、核实和核证温室气体（GHG）源人为减排量，以及向碳交易机制理事会提出申请审核 CERs 的独立实体。每个指定经营实体都是仅为某种碳交易活动（从事的部门范围）而得到授权的，有可能是国有部门或国际机构。

85. 核证（certification），由指定的经营实体（DOE）提出的书面保证，即在一个具体时期内某项目活动所实现的温室气体源人为减排量已被核实。

86. 核证减排量（certified emission reductions，CER），指一单位符合碳交易机制原则及要求，且经联合国执行理事会（EB）签发的交易机制或 PoAs（活动规划类）项目的减排量，一单位 CER 等同于一吨的二氧化碳当量，计算 CER 时采用全球变暖潜力系数（GWP）值，把非二氧化碳气体的温室效应转化为等同效应的二氧化碳量。

87. 国家核证自愿减排量（Chinese certified emission reduction，CCER），指我国依据国家发展和改革委员会发布施行的《温室气体自愿减排交易管理暂行办法》的规定，经其备案并在国家注册登记系统中登记的温室气体自愿减排量。

88. 计入期（crediting period），指项目情景相对于基线情景产生额外的温室气体减排量的时间区间。项目参与者应当将计入期起始日期选定在自愿减排项目活动产生首次减排量的日期之后，计入期不应当超出该项目活动的运行周期。项目参与方可选择固定计入期或可更新计入期两种。

89. 固定计入期（fixed crediting period），也称"计入期—固定"（also crediting period-fixed），是用来确定计入期期限的两个备选方案之一。在该方案中，项目活动的减排额计入期期限和起始日期只能一次性确定，即一旦该项目活动完成登记后不能更新或延长。

90. 可更新计入期（renewable crediting period），也称"计入期—可更新"（also crediting period-renewable），是用来确定计入期期限的备选方案之一。在该方案中，林业碳汇项目可为 20

年。这一计入期最多可更新两次，即最大为 60 年。

91. 注册（registration），指执行理事会（ED）或国家规定的管理机构正式接受一个经确认合格的项目活动为一项碳汇交易机制项目活动。注册是核实、核证及颁发与这一项目活动相关的经核证排减量（CERs）或国家核证减排量（CCER）的先决条件。

92. 核实（verification），指由指定的经营实体（DOE）定期独立审评和事后确定已登记的碳交易机制项目活动在核实期内产生的、经监测的温室气体（GHG）源人为减排量。

93. 监测（monitoring），指收集和归档所有对确定基准线，测量某一减排项目（CDM 或自愿减排项目）活动在项目边界内的温室气体（GHG）源人为排放量以及泄漏所必要的并可适用的相关数据。

94. 基线碳汇量（baseline net green house gas removal by sinks），也称"基线净温室气体汇清除"，是基线情景下项目边界内各碳库中的碳储量变化之代数和。

95. 项目碳汇量（actual net green house gas removal by sinks），也称"实际净温室气体汇清除"，是项目情景下项目边界内所选碳库中的碳储量变化量，减去由碳汇造林项目活动引起的项目边界内温室气体排放的增加量。

96. 项目减排量（net anthropogenic green house gas removal by sinks），也称"净人为温室气体汇清除"，指由于造林项目活动产生的净碳汇量。项目减排量等于项目碳汇量减去基线碳汇量，再减去泄漏量。

97. 透明和保守的（transparent and conservative），以透明和保守的方式确定基线所做的假设条件，并且所作的选择是可证实的。在变量和参数值的不确定情景时，对基线所作的预测不过高估计某一碳交易机制项目活动的减排量，则该基线的确定即被视为是偏保守的。

98. 重点排放单位（key emission unit），指满足国务院碳交易主管部门确定的纳入碳排放权交易标准且具有独立法人资格的温室气体排放单位。

99. 清单（inventory），机构的温室气体排放量和排放源的量化表。

100. 排放因子（emission factor），量化每单位活动的气体排放量或清除量的系数。排放因子通常在给定的一组操作条件下，基于测量样本数据得到具有代表性的平均活动水平的排放率。

101. 农地（cropland），包括可耕地和耕地，以及农林系统中植被低于林地阈值且与国家选择的定义相一致的土地。

102. 农田管理（cropland management），种植农作物的土地和休耕的或暂时不用于作物生产的土地做法体系。

103. 决策树（decision tree），决策树是一个描述具体规定步骤的流程图。在依据优良作法原则编制清单或清单分量时，需要按此顺序进行。

104. 干扰（disturbances），减少或重新分配陆地生态系统碳库的过程。

105. 专家评价（expert judgment），指经过仔细审议且翔实记录的定性或定量评价，这些评价由一个或数个有特定领域专门技能的人员在没有不含糊观测证据情况下作出。

106. 措施（measure），在气候政策中，措施是促进气候变化减缓的技术、流程和做法。例如可再生能源技术、废弃物最少化流程以及公共交通做法等。

107. 草地（grassland），这一类别包括牧场和不被视为农田的牧草地。它还包括植被未能达到林地标准的且通过人为干预未能达到林地标准的类别。草地还包括从荒地到休闲区的所有草地，以及农业和林牧系统，分为管理和非管理两类，同国家定义一致。

108. 湿地（wetlands），这一类型包括全年或一年部分时间被水覆盖或处于水饱和状态且不属于林地、耕地、草地或定居地类型的土地（例如泥炭地）。这一类型可按通用定义细分为管理和非管理类型，即它包括属于管理分类的水库和属于非管理分类的天然河流和湖泊。

109. 聚居地（settlement），这一类别包括所有开发的土地，包括任何规模的运输基础设施和人类居住地，除非它们已被列入其他类别之下。这应与国家选择的定义相一致。

110. 其他土地，作为一种土地利用类型（other lands，a land-use category），这一类型包括裸土、岩石、冰和所有不属于任何其他 5 种类型的未管理的土地。在可获得数据的情况下，它允许认定的土地总面积与全国面积匹配。

111. 牧场管理（grazing land management），指用于畜牧生产的土地上旨在调控所产饲料和牲畜的数量和类型的一套做法。

112. 全年毛总增量（gross annual increment），参考期内按规定的最小胸径（各国不同）测量的所有树木的蓄积平均年增量。包括已被采伐或死亡的树木的增量。

113. 定义的统一（harmonization of definition），在这一背景下指使定义标准化或增强定义间的可比性和一致性。

114. 可比性（comparability），指缔约方报告的清单中排放和清除估算应该在缔约方间进行比较。为此，缔约方应该使用缔约方大会（COP）通过的方法和格式来进行估算和报告清单。

115. 相关性（relevance），确保温室气体排放清单恰当地反映企业的温室气体排放情况，服务于企业内部和外部用 户的决策需要。

116. 完整性（completeness），指清单包括全地域覆盖的所有源和汇以及包括《1996 年 IPCC 国家温室气体清单指南修订本》中的所有气体，还包括个别缔约方特定的其他有关的源 / 汇类别。

117. 一致性（consistency），指清单在数年时间范围内其所有要素应该内在一致。如果对基准年和所有其后年份使用同一方法，使用一致的数据集估算源排放或汇清除，那么清单是一致的。

118. 透明性（transparency），指报告编制者应该清楚地解释清单所用的假定和方法，披露任何有关的假定，并恰当指明所引用的核算与计算方法学以及数据来源，以促使清单编制的重复性和评估清单。

119. 基准年（base year），清单的起始年。目前一般是以 1990 年为基准年。

120. 活动（activity），在给定的时期和界定的区域内所发生的一项作业或一系列作业。

121. 活动数据（activity data），在一定的时间内引起温室气体源排放或清除的人类活动数量的大小。在土地利用、土地利用变化和林业（LULUCF）部门，土地面积、蓄积、经营管理系统、石灰和肥料的使用等 数据均是活动数据的例子。

122. 关键类别（key category），指在国家清单体系内占有优先位置的类别，无论是其排放绝对值还是排放趋势的估算或这两个方面都对国家温室气体排放清单总量产生重要的影响。

123. 土地覆盖（land cover），土地表面覆盖的植被类型。

124. 土地利用（land use），清单编制中在一个土地单位上开展的活动类型。在《土地利用、土地利用变化和林业方面的优良 做法指南》（*Good Practice Guidance for Land Use, Land-Use Change and Forestry*，GPG-LULUCF）中，这一术语用于界定土地利用 类别，这些土地类别是土地覆盖（例如森林、草地、湿地）和土地利用（例如耕地、定居地）种类的混合体。

125. 土地利用、土地利用变化和林业（land use,land use change and forestry, LULUCF），是国家温室气体清单报告的一个部分，涵盖直接由人类引起的土地利用、土地利用变化 和林业活动带来的温室气体排放和清除，不包括农业排放。

126. 农业、林业和其他行业土地利用（agriculture, forestry and other land use, AFOLU），是《2006 年 IPCC 国家温室气体清单指南》中明确的规定清单报告内容，涵盖直接由人类引起的土地利用、土地利用变化和林业活动带来的温室气体排放和清除。AFOLU 与 LULUCF 相比，避免了部分活动导致的温室气体交叉问题。

127. 减少毁林和森林退化所致排放量（reducing emissions from deforestation and forest degrada-

tion，REDD），是为发展中国家提供激励措施，使其减少因森林破坏等导致的碳排放，从而为储存在森林中的碳创造金融价值的做法，所以也是一种通过避免毁林实现减缓的机制。REDD+ 比再造林和森林退化更为广泛，包括森林保护和可持续管理及加强森林碳储存的作用。这个概念第一次提出是在 2005 年蒙特利尔召开的 UNFCCC 第 11 次缔约方大会（COP）上，2007 年在巴厘岛召开的 UNFCCC 第 13 次 COP 高度承认了这个概念并将其纳入了"巴厘岛行动计划"，呼吁建立"关于减少发展中国家毁林和森林退化所致排放量（REDD）和发展中国家森林保护、森林可持续管理及加强森林碳储存作用相关活动的政策方法和积极的激励措施（REDD+）"。

128. 经营草地（managed grassland），在其上开展人为活动的草地，如放牧或收割干草等活动。

129. 年净增量（net annual increment），在给定参考期内，按规定的最小胸径测量的所有树木的总增量减去自然死亡量的年平均量。

130. 净—净核算（net-net accounting），报告年的碳汇或碳源减去基准年的碳汇或碳源。《京都议定书》第 3、4 条规定了放牧地管理、耕地管理和植被恢复的核算法。

131. 总—净核算（gross-net accounting），该方法根据项目经营管理的碳汇增长率和消耗率的比例直接计算报告年的净碳汇或碳源。该方法一般比净—净核算法结果大。

132. 有机土壤（organic soil），符合以下 Ⅰ 和 Ⅱ 或 Ⅰ 和 Ⅲ 所列要求的土壤为有机质土壤（粮农组织，1998）；厚度为 10 cm 或以上。当混合深度为 20 cm 时，小于 20 cm 厚的一层必须有 12% 或更多 的有机碳；如果土壤几天仍没有达到饱和水，而且有机碳（大约 35% 的有机质）含量超过 20%（按 重量）；如果土壤常处于水分饱和情形并符合下列任一条件：（Ⅰ）如果无黏粒，有机碳至少为 12%（按重量）（约 20% 的有机质）；（Ⅱ）如果黏粒含量在 60% 或以上，有机碳至少为 18%（按重量）（约 30% 的有机质）；或 （Ⅲ）介于二者之间，中间量的黏土有成比例的有机碳量。

133. 泥炭土（peat soil），也称"有机土"（histosol），一种典型的湿地土壤，水位高而且有机质层至少 40 厘米厚（排水不畅的有机质土）。

134. 沙质土（sandy soil），包括砂粒含量超过 70% 和黏粒含量低于 8% 的所有土壤（不管分类如何）（基于标准质地测量（粮农组织分类包括砂土、砂质岩成土））。

135. 做法（practice），对土地、与土地有关的碳库储量或对温室气体与大气的交换产生影响的一项或一组行动。

136. 优良做法（good practice），优良做法是一套规范，目的是确保温室气体清单准确性，即在当前判断能力情况下既不过高也不过低估算碳排放，而且从实际操作方面尽可能地减少不确定性。优良做法包括选择适合国家实际情况的估算方法、国家层面的质量保证和质量控制、不确定性的量化以及有利于提高透明度的资料存档与报告。

137. 报告（reporting），向《联合国气候变化框架公约》提供估算国家温室气体清单的过程。

138. 恢复力（resilience），某社会、经济和环境系统处理灾害性事件、趋势或扰动，并响应或重组，同时保持其必要功能、定位及结构，并保持其适应、学习和改造等能力的能力。

139. 分辨率（resolution），可以确定有关土地覆盖或利用情况的最小土地单位。高分辨率指可分辨的土地单位小。

140. 植被恢复（revegetation），在有关地点通过建立覆盖面积至少为 0.05 公顷的植被以增加碳储量，而且不满足形成森林定义的另一种直接人为活动。

（文献来源：泰达低碳中心.140 个碳中和专业术语.https://mp.weixin.qq.com/s/XZAfZl40wSFlkDaDCw6eHg；知乎用户奶爸双碳研究所.140 个碳中和专业术语.https://zhuanlan.zhihu.com/p/453901938）